SOCIETY FOR EXPERIMENTAL BIOLOGY
seminar series: 26

THE CELL DIVISION CYCLE IN PLANTS

The cell division cycle in plants

Edited by
J. A. BRYANT
Reader in Plant Biochemistry,
Department of Plant Science,
University College, Cardiff
D. FRANCIS
Lecturer in Cell Biology,
Department of Plant Science,
University College, Cardiff

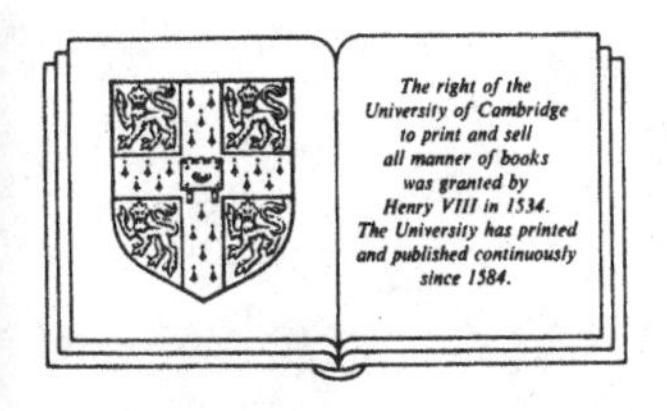

CAMBRIDGE UNIVERSITY PRESS
Cambridge
London New York New Rochelle
Melbourne Sydney

CAMBRIDGE UNIVERSITY PRESS
Cambridge, New York, Melbourne, Madrid, Cape Town, Singapore, São Paulo, Delhi

Cambridge University Press
The Edinburgh Building, Cambridge CB2 8RU, UK

Published in the United States of America by Cambridge University Press, New York

www.cambridge.org
Information on this title: www.cambridge.org/9780521103619

First published 1985
This digitally printed version 2009

A catalogue record for this publication is available from the British Library

Library of Congress Catalogue Card Number: 84-46159

ISBN 978-0-521-30046-9 hardback
ISBN 978-0-521-10361-9 paperback

CONTENTS

CONTRIBUTORS

Bayliss, M.
Corporate Bioscience and Colloid Laboratory, Imperial Chemical Industries PLC, Runcorn, Cheshire, WA7 4QE, U.K.

Bennett, M.D.
Cytogenetics Department, Plant Breeding Institute, Maris Lane, Trumpington, Cambridge CB2 2LQ, U.K.

Boffey, S.A.
Division of Biological and Environmental Sciences, The Hatfield Polytechnic, P.O. Box 109, Hatfield, Herts, U.K.

Bryant, J.A.
Department of Plant Science, University College, P.O. Box 78, Cardiff, CF1 1XL, U.K.

Cavalier-Smith, T.
Department of Biophysics, King's College London, 26-29 Drury Lane, London WC2B 5RL, U.K.

Clayton, L.
Department of Cell Biology, John Innes Institute, Colney Lane, Norwich, NR4 7UH, U.K.

Dunham, V.L.
Department of Biology, State University College, Fredonia, New York 14063, U.S.A.

Francis, D.
Department of Plant Science, University College, P.O. Box 78, Cardiff CF1 1XL, U.K.

Kelly, P.
Department of Botany, Edinburgh University, Mayfield Road, Edinburgh EH9 3JH, U.K.

Kidd, A.D.
Department of Plant Science, University College, P.O. Box 78, Cardiff, CF1 1XL, U.K.

Lyndon, R.F.
Department of Botany, Edinburgh University, Mayfield Road, Edinburgh, EH9 3JH, U.K.

Nagl, W.
Abteilung Zellbiologie, Fachbereich Biologie der Universität Kaiserslautern, Postfach 3049, D-6750 Kaiserslautern, Federal Republic of Germany.

Pohl, J.
Abteilung Zellbiologie, Fachbereich Biologie der Universität Kaiserslautern, Postfach 3049, D-6750 Kaiserslautern, Federal Republic of Germany.

Radler, A.
Abteilung Zellbiologie, Fachbereich Biologie der Universität Kaiserslautern, Postfach 3049, D-6750 Kaiserslautern, Federal Republic of Germany.

Scherthan, H.
Abteilung Zellbiologie, Fachbereich Biologie der Universität Kaiserslautern, Postfach 3049, D-6750 Kaiswerslautern, Federal Republic of Germany.

Shall, S.
Cell and Molecular Biology Laboratory, Biology Building, University of Sussex, Brighton, BN1 9QG, U.K.

Trewavas, A.J.
Department of Botany, Edinburgh University, Mayfield Road, Edinburgh, EH9 3JH, U.K.

Van't Hof, J.
Department of Biology, Brookhaven National Laboratory, Upton, New York 11973, U.S.A.

Waterborg, J.H.
Cell and Molecular Biology Laboratory, Biology Building, University of Sussex, Brighton, BN1 9QG, U.K.

PREFACE

This volume is the published proceedings of a Seminar Series Symposium on the Cell Division Cycle in Plants which formed part of the 213th meeting of the SEB, held in April 1983 at Cardiff. In meetings on the Cell Cycle held in recent years, virtually all the topics covered have been from prokaryote, yeast or mammalian systems. We therefore considered that the time was ripe for an in-depth treatment of the plant cell cycle.

The initial part of the meeting concentrated on DNA replication, chromatin organisation, the mitotic apparatus, and the various biochemical steps at which control of the cell cycle could be mediated. Then followed a series of papers concerned with aspects of cell cycle control, ranging from the molecular, through the cellular to the organ level, the latter being examplified by the pre-floral shoot meristem. Finally, two more specialised topics, namely DNA endoreduplication cycles and chloroplast division were dealt with.

One of our intentions in organising this meeting was to bring together biochemists, molecular and cell biologists and plant physiologists in order to take a multidisciplinary approach to the cell cycle. Our invited speakers, and the many others who contributed to discussion, enthusiastically took up this approach, and we are very grateful to the contributors for that. We can only hope that this volume in some way reflects the lively and enthusiastic atmosphere of the meeting.

In addition to thanking the contributors, we also wish to thank the many others who have assisted us in various ways, especially Unilever Research PLC and I.C.I. PLC for financially supporting the meeting, Mrs. Hilary Webb for typing, the staff of CUP for their help and encouragement, and finally, Dr. S.W. Armstrong, Mr. A.D. Kidd, Mr. J.C. Ormrod and Dr. D.R. Sibson for their assistance with proof-reading (although, of course we accept full responsibility for the final product).

John A. Bryant
Dennis Francis

Editors for the Society
of Experimental Biology

CONTROL POINTS WITHIN THE CELL CYCLE

J. Van't Hof

INTRODUCTION

More than thirty years ago Howard and Pelc (1953) recognized that the interphase of dividing cells of *Vicia faba* had three phases. They named these phases the presynthetic phase (G1), the DNA-synthetic phase (S), and the postsynthetic phase (G2). Cells in G2 enter mitosis (M) and after division the daughter cells enter G1. The term cell cycle refers to the passage of cells in time from one division to the next, starting at M, proceeding to G1, on to S, then to G2, and finally to M again.

In root meristematic cells of unrelated species the duration of S is determined by the amount of nuclear DNA (Van't Hof, 1965, 1974a). Cell division and cell differentiation, on the other hand, are controlled by factors that operate during G1 and G2 (Van't Hof & Kovacs, 1972; Van't Hof, 1973, 1974b). These two observations constitute the general framework and the motivation for much of the material covered under the four topics presented in this chapter. The first topic, the principal control point hypothesis and the control of cell division in plants from G1 and G2 is reviewed briefly. The second concerns the positive relation between nuclear DNA content and the duration of S. An interpretation of this relation is presented in terms of replicon properties. Third, chromosomal DNA maturation, its occurrence in late S, and its dependence on high cellular concentrations of thymidine are mentioned. Finally, the presence of extrachromosomal DNA in certain cells that differentiate from G2 is discussed.

PRINCIPAL CONTROL POINT HYPOTHESIS

The idea that cell division in complex tissues reflects the activity of genetically different cell populations can be traced to Gelfant (1961, 1963, 1966). Cell populations were identified phenotypically by the phase in which they arrest when not dividing; one population arrested in G1, the other in G2. Meristematic cells likewise arrest in G1 and G2 when they cease dividing (Van't Hof, 1973, 1974b). The reproducibility of this phenomenon, whether induced *in vitro* by nutritional starvation or *in vivo* by dormancy, was the basis for the principal control point hypothesis (Van't Hof & Kovacs, 1972). The hypothesis states that cell division in complex tissue is regulated by factors that operate during G1 and G2 and that under conditions where cell division ceases these factors become limiting, causing cell arrest in G1 or G2. Evidence supporting the hypothesis is extensive and the topic has been comprehensively reviewed (Rost, 1977). Though the hypothesis was based on observations made on cells of complex tissue, recent work with plant cell suspension cultures shows it to be valid also for single cells (Gould, Everett, Wang & Street, 1981). The principal controls are manifested at two stages during the lifetime of root cells; once when the cells are proliferative but cease dividing temporarily because of stress or other causes and again when they cease dividing and differentiate to form mature tissue (Evans & Van't Hof, 1974a). Cytological data show that the phase in which cells of higher plants arrest is prescribed, suggesting that the phenomenon is genetically controlled. The ratio of cells arrested in G1 : G2 in nutritionally starved root meristems is species-specific and similar ratios are seen in mature root tissue of unstressed seedlings (Evans & Van't Hof, 1974a). This finding demonstrates that the phase in which cells arrest (arrested cell phenotype) is consistent and independent, regardless of the cause of arrest. The presence of genes in yeast that control cell division in G1 and G2 augments the possibility that similar genes may be functioning in higher plant cells (Beach, Durkacz & Nurse, 1983).

Though meristematic cells arrest only in G1 and G2, this does not mean that the meristem has only two cell populations. Pea root meristems have at least three cell populations, each with its own arrested cell phenotype (Evans & Van't Hof, 1974b). One population arrests in G1, another in G2, and a third stops in either G1 or G2. This latter group is responsive to a substance called the G2 factor (Evans & Van't Hof, 1973, 1974b) now identified as 5-methylnicotinic acid or trigonelline (Evans, Almeida, Lynn & Nakanishi, 1979). The cells arrest in G2 when trigonelline is present continuously or in G1 if it is absent. Most legumes have high concentrations of trigonelline (71 to 554 μg/g) while most non-legumes and monocotyledonous plants have 0.6 to 15 μg/g of tissue (Evans & Tramontano, 1984). Work with trigonelline is still in the descriptive stage and its direct mechanism of action on cells remains unknown. Nevertheless, trigonelline appears to have all the classical characteristics of a plant hormone (Evans & Tramontano, 1981).

The principal control point hypothesis originated from observations made on root meristem cells but experiments with shoot apices indicate that cell division in this tissue also is regulated in G1 and G2. In 1967, Bernier and his associates (Bernier, Kinet & Bronchart, 1967) suggested that the transition from a vegetative to a floral meristem involved the accumulation of cells in G2 prior to increased mitotic activity. This suggestion was confirmed later by cytophotometric measurements (Jacqmard & Miksche, 1971). That cells first accumulate in G2 after a change in photoperiod argues for a light-sensitive controlling factor that operates in G2. Also, in the shoot apex of *Silene coeli-rosa* only G1 and G2 respond to far-red light and red light exposures, again indicating that the cell cycle is controlled by factors that operate during these phases (Francis, 1981a). Finally, a comparison of the number of cells in G1 and G2 (G1:G2 ratio) in the vegetative apex of three plants shows that the ratio is species specific just as in root meristems. *Sinapis alba* has a ratio favouring G2 (Jacqmard & Miksche, 1971), *Xanthium strumarium* favours G1 (Jacqmard, Raju, Kinet & Bernier, 1976), and *Silene coeli-rosa* has a ratio of almost one (Francis & Lyndon, 1978; Francis, 1981b). These data are consistent with the principal control point hypothesis and they indicate that G1 and G2 are the phases where the initial photo-responses are registered in the shoot apex (see Francis & Lyndon, this volume).

Observations consistent with a hypothesis, however, are not enough. Cellular work on root and shoot meristems alike remain limited by the lack of information at the molecular level. The controls that operate during G1 and G2 need to be defined in molecular terms and more effort must be directed toward the question of why G1 and G2 exist at all. It may be more productive to adopt the view that the entire cell cycle, not just S and M, is a mechanism to assure a correct and adequate genetic inheritance to each of the recipient daughter cells produced by mitosis. This point of view is advantageous because it focuses on chromatin and DNA, and as such, deals directly with the most important elements of molecular controls.

REPLICON PROPERTIES AND THE DURATION OF S

The duration of S is the time needed by a cell to replicate its chromosomes. In cellular terms, the amount of DNA in a haploid nucleus (genome size) is defined as a *C*-value. When diploid cells begin S, they have a 2*C* amount of DNA and when they end S, they have a 4*C* amount. Chromosomal DNA in eukaryotic cells is replicated by numerous replication units called replicons (Taylor, 1963; Huberman & Riggs, 1968). Each replicon has an origin where replication begins and two replication forks that diverge bi-directionally from the origin while replicating nascent DNA chains. Two properties of replicons, size (the origin to origin distance) and the rate of fork movement, vary little amongst plant species with genome sizes that differ 82-fold (Van't Hof & Bjerknes, 1981). Taken as a group these plants have replicons with a mean size of 22 ±3.4 μm and a mean fork rate of 8 ±1.4 μm per hour at 23°C.

A common feature of chromosomal DNA replication is that replicons function in groups (Cairns, 1966; Huberman & Riggs, 1968; Hand, 1975). Along a single chromosomal duplex molecule, several tandem replicons actively replicate DNA while a neighbouring group located on the same molecule remains inactive (Fig. 1). This feature suggests that chromosomal DNA replication is organized as a three-unit hierarchy. The elementary replication unit is a single replicon which is a member of a group of replicons arranged end to end along a section of the chromosomal duplex. Such a group, called a cluster is distinguished by the fact that its members replicate their portion of nascent DNA almost simultaneously. The third unit of the hierarchy is called a family or bank. It consists of many clusters that are distributed amongst

Fig. 1. An autoradiogram of chromosomal DNA fibres of *Arabidopsis thaliana* isolated after a 45 minute pulse with tritiated thymidine. The bar scale is 500 μm and the arrows point toward gaps of unlabelled DNA that accommodate one or more replicons, i.e. gaps that are 24 μm or longer. The tandem arrays of silver grains trace the movement of replication forks during the pulse. Note that replicons on the same DNA fibre replicate in clusters. The clusters are separated by gaps of unlabelled DNA; these gaps are occupied by replicons that replicate DNA either before or after the pulse. (from Van't Hof, Kuniyuki & Bjerknes, 1978).

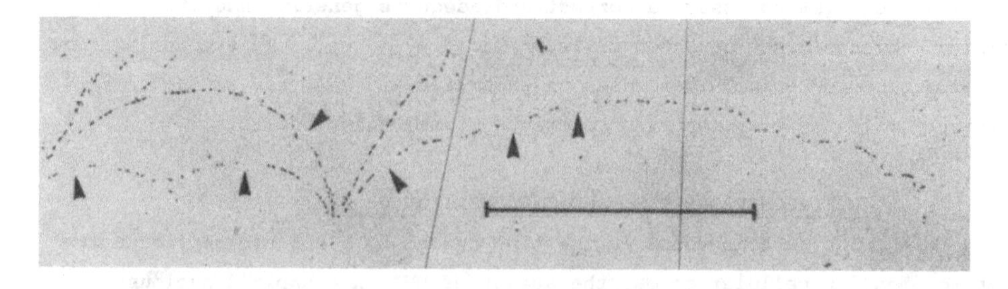

the chromosomal complement of the cell. A family is operationally defined as one or more clusters that replicate at a certain time during S. The temporal order of DNA replication during S is then, a reflection of the sequential activity of replicon families. An example of sequential activity of replicon families was detected in dividing cells of *Arabidopsis thaliana* (Van't Hof, Kuniyuki & Bjerknes, 1978). *A. thaliana* has two replicon families, one estimated to have 687 members and another with 1888 members per genome. The families initiate replication in sequence separated by a 36 minute interval. The S-phase in *A. thaliana* is about 2.8 hours and 95% of this time is accountable by two variables; the time needed by each replicon family to replicate its portion of chromosomal DNA and the 36 minute interval between the beginning of replication by the first family activated when cells enter S and the second family activated 36 minutes later. The diagram in Fig. 2 shows how this temporal relation is viewed for *A. thaliana*. It is proposed that cells of other plant species with larger genomes have more replicon families, and correspondingly longer S-phases (Van't Hof & Bjerknes, 1981; see also Francis, Kidd & Bennett, this volume.)

The sequential activity of replicon families during S suggests that factors exist that are responsible for maintaining the order of replication and it implies that these factors are specific for each replicon family. The nature of these factors and how they are generated is unknown but it is obvious that those responsible for activation of the first family to replicate in S are produced during G1. Factors specific for the activation of the second family may, in turn, be products of genes encoded on replicons of the first family. If transcription of these genes is delayed until after their

Fig. 2. A diagram showing the temporal order of replication of the two replicon families of *Arabidopsis thaliana* during S. Family A is the first to replicate, family B begins replication about 36 minutes later. The combined time for both families to replicate their DNA plus the 36 minutes interval comprises 95% of S.

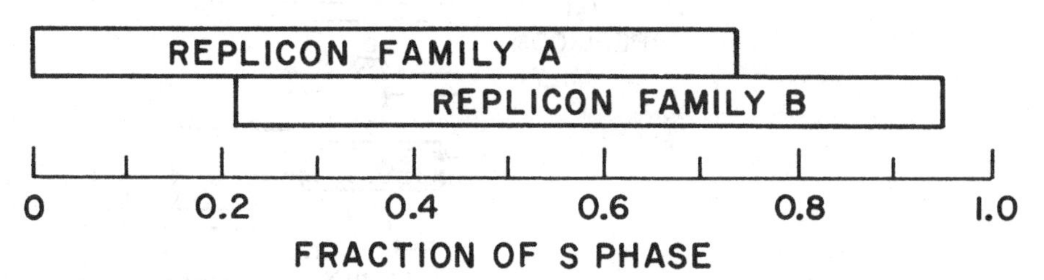

sequences are replicated, then the time of their expression is dependent on the replication fork rate. The faster the forks move, the sooner these genes are replicated and transcribed, and the sooner the next replicon family is signalled to begin replication. In cells with several replicon families, this type of auto-regulation would produce a cascading effect at the replicon level until cells complete S (Fig. 3).

The idea that the temporal order of replicon families is self-regulating once the first family begins is supported, but not proven, by measurements of replicon properties of meristematic cells of *Helianthus annuus* roots grown at different temperatures (Van't Hof, Bjerknes & Clinton, 1978). At temperatures from 20 to 35°, a change in replication fork rate

Fig. 3. Model for the auto-regulation of sequential replication of replicon families in plant cells during S phase. A factor produced during G1 activates the 1st family to replicate when the cell enters S. Replication begins at the origin, noted as O, and nascent chain growth proceeds bidirectionally replicating gene a. Gene a is immediately transcribed and its gene product activates the origins of the 2nd family. Gene b is subsequently replicated, transcribed, and its product activates origins of the 3rd family, etc. Note that the faster the replication forks move, the sooner genes a, b, and c are replicated and the shorter the duration of S. Parental DNA molecules are represented by the solid lines, nascent chains by the dashed lines, and replication forks by the arrows.

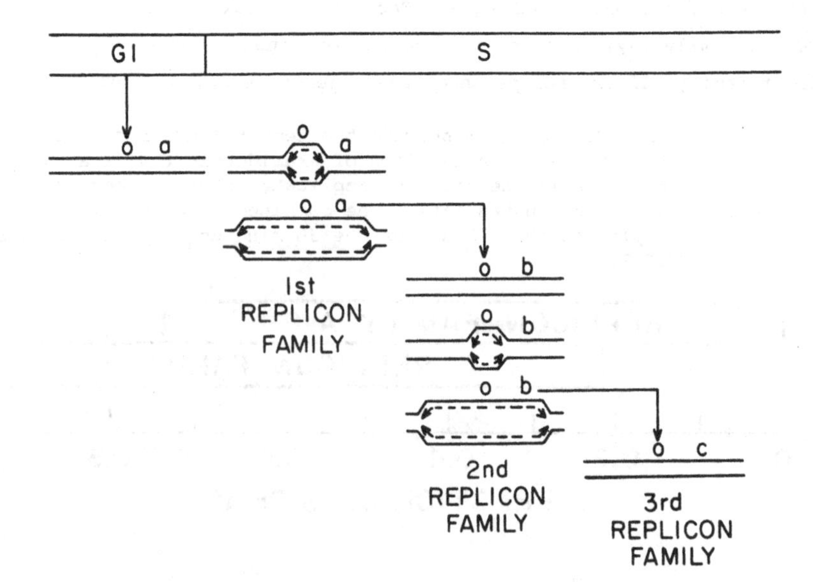

accounts for more than 90% of the change in the duration of S. Thus, at these temperatures the time between the replication of one family and that of the next depends on the replication fork rate. Only at 38°, a near lethal temperature for H. annuus, or at low temperatures (10 to 15°) is the lengthening of S, and hence, the interval between the replication of one family and the next, due to processes other than fork rate.

THYMIDINE CONCENTRATION EFFECTS ON CHROMOSOMAL DNA MATURATION

Chromosomal DNA maturation is unique to eukaryotic cells. The term maturation refers to the joining of nascent replicons to form cluster-sized molecules and to the joining of clusters to produce chromosomal-sized nascent DNA. In plant and animal cells chromosomal DNA maturation occurs in a stepwise manner that is cell cycle dependent (Kowalski & Cheevers, 1976; Walters, Tobey & Hildebrand, 1976; Funderud, Andreassen & Haugli, 1978; Schvartzman, Chenet, Bjerknes & Van't Hof, 1981). The first step (Step I, Fig. 4) occurs in early S with the initiation, chain elongation, and termination of nascent chains by replicon clusters. The replicons of pea root cells need about an hour to replicate a nascent single-stranded chain of about 18 x 10^6 daltons (Van't Hof & Bjerknes, 1977; Van't Hof, Bjerknes & Delihas, 1983). After this, the replication forks of adjacent replicons within a cluster converge and stop, leaving a gap between nascent chains (Step II). Meanwhile the neighbouring cluster, located to the right of the first in the diagram of Fig. 4, began replication. As cells progress through S, each cluster of each replicon family follows Steps I and II until most of the chromosomal DNA is replicated with the exception of the gaps between adjacent nascent replicons. When cells reach late S or early G2 the gaps between nascent replicons are sealed, first producing cluster-sized molecules (Step III) and then chromosomal-sized nascent chains of DNA complementary to the parental strand (Step IV). Recent work with pea roots indicates that the amount of thymidine available to cells in S is a factor responsible for the stepwise maturation of chromosomal DNA (Schvartzman, Krimer & Van't Hof, 1984). Synchronized cells starved of carbohydrate and thymidine fail to join nascent replicons when fed thymidine exogenously at concentrations of 1 to 10 μM but do so when the concentration is 100 μM. Thus, the final step of maturation occurs when the relative concentration of thymidine is high. This observation is in accord with an idea proposed by Mathews and Sinha (1982). Using T4 phage-infected Escherichia coli, these authors determined that the concentration of deoxythymidine triphosphate was 4 to 5 times higher at the replication fork than in other areas within the cell. They proposed the existence of a concentration gradient near the replication fork that results in less thymidine being available for other processes. For the eukaryotic cell one of these other processes would be chromosomal DNA maturation. Thus, in replicating pea root cells, an exogenous supply of 100 μM thymidine is sufficient to overcome or reduce an effect of a concentration gradient at the replication forks and is high enough for cells to join nascent replicons. The idea that chromosomal DNA maturation occurs when the relative cellular pool of thymidine is high, is supported by other work with plant and mammalian cells. Chromosomal DNA maturation in mammalian cells occurs in late S (Kowalski & Cheevers, 1976; Tobey & Hildebrand, 1976), a time when the cellular concentrations of deoxyribonucleoside triphosphates, and particularly deoxythymidine tri-

Fig. 4. A diagram showing the stepwise replication and maturation of chromosomal DNA during S and G2. Top, the parental chromosomal duplex in G1 with O's representing replicon origins; Step 1, bidirectional replication by four replicons in a cluster; Step II, the convergence of replication forks of neighbouring replicons within the cluster leaving a single-stranded gap between the nascent chains; to the right of the first cluster a second has begun replication; Step III, gaps between nascent chains of replicons of the first cluster are sealed and joined producing a cluster-sized molecule; Step IV, gaps between neighbouring clusters and replicons are sealed and joined to give chromosomal-sized DNA. Note that the time between the replication of the first and that of the second cluster may be greater than shown.

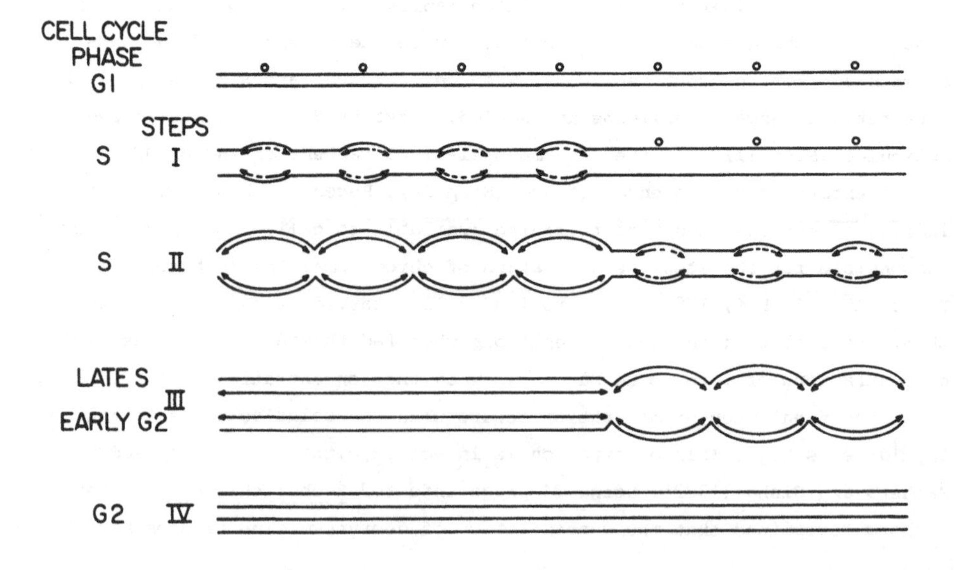

phosphate, are highest (Skoog, Nordenskjold & Bjursell, 1973; Walters, Tobey & Ratliff, 1973). Also, in plant cells two enzymes, deoxythymidine kinase and deoxythymidine monophosphate kinase, are more active during late S and early G2 than in early S (Harland, Jackson & Yeoman, 1973). These enzymes provide precursors for deoxythymidine triphosphate, and their high activity during late S and early G2 would increase the cellular pool of deoxythymidine triphosphate, favouring chromosomal DNA maturation.

EXTRACHROMOSOMAL DNA AND CELL DIFFERENTIATION FROM G2

The root tip meristem serves the plant in two ways; it is the site of cell proliferation, hence responsible for the building blocks of root growth, and it is the site where cells take their first step toward differentiation. When cells cease dividing they remain arrested in G1 or G2 and begin differentiation. In pea roots certain cells in their final cell cycle replicate about 80% of their nuclear DNA, accumulate in late S, replicate the remaining 20% of their DNA, and eventually differentiate from G2 (Van't Hof & Bjerknes, 1982; Van't Hof, Bjerknes & Delihas, 1983). When replicating the remaining 20% of their DNA, these cells produce extra-chromosomal DNA molecules (exDNA). Three pieces of evidence indicate that exDNA is of nuclear origin. First, more than 80% of selectively extracted exDNA molecules have a buoyant density that is the same as nuclear DNA. Second, the molecules are linear and not circular as is expected of plastid or mitochondrial DNA (Krimer & Van't Hof, 1983). Third, the molecules of exDNA have methylated bases as does nuclear but not organelle DNA (J. Van't Hof, unpublished results).

Extracted exDNA molecules are replicon size (54 to 73 kilobase pairs) suggesting that they may be free replicons produced either by amplification or by excision from the chromosomal duplex during late S (J. Van't Hof, unpublished results). Once free of the chromosome, portions of the exDNA duplex molecule are replicated by a strand displacement mechanism producing single-stranded DNA molecules of about 11 kilobases (Krimer & Van't Hof, 1983). Single-stranded exDNA is found in wheat also and these molecules are postulated to be involved with cell differentiation (Buchowicz, Kraszewska & Eberhardt, 1978; Kraszewska & Buchowicz, 1983).

In pea, there is cytological evidence that cells producing exDNA differentiate from G2 and form epidermal and stelar tissues (J. Van't Hof, unpublished results). While not definitive, this observation suggests a relation between exDNA and cells that differentiate to form certain tissues within the root.

The function of exDNA is unknown, but it is clear that its production is one of the first steps in the transition of cells from a proliferative to a differentiated state. One working hypothesis is that exDNA molecules are late replicating replicons that are no longer attached to the chromosomal duplex and that these replicons have genes whose products are needed for the differentiation of certain cell types. That a portion of the exDNA duplex molecule is removed by strand displacement is curious, and there are two possible explanations for this phenomenon. First, it is possible that strand displacement removes certain base sequences that remained modified, and presumably unexpressed, when the cells were meristematic. The sequences replacing those removed may be unmodified and therefore expressed while cells differentiate. A second possibility is that the production of free single-stranded molecules is a means of amplifying certain base sequences needed for cell differentiation.

SUMMARY

The diagram shown in Fig. 5 outlines the cell cycle and notes the temporal relation between the topics presented in the text. Several details are excluded but these are mentioned elsewhere (Van't Hof & Kovacs, 1972; Van't Hof, 1973, 1974b; Rost, 1977). Also absent are details about

Fig.5. A diagram of the cell cycle in higher plants showing the relative positions of events discussed in the text.

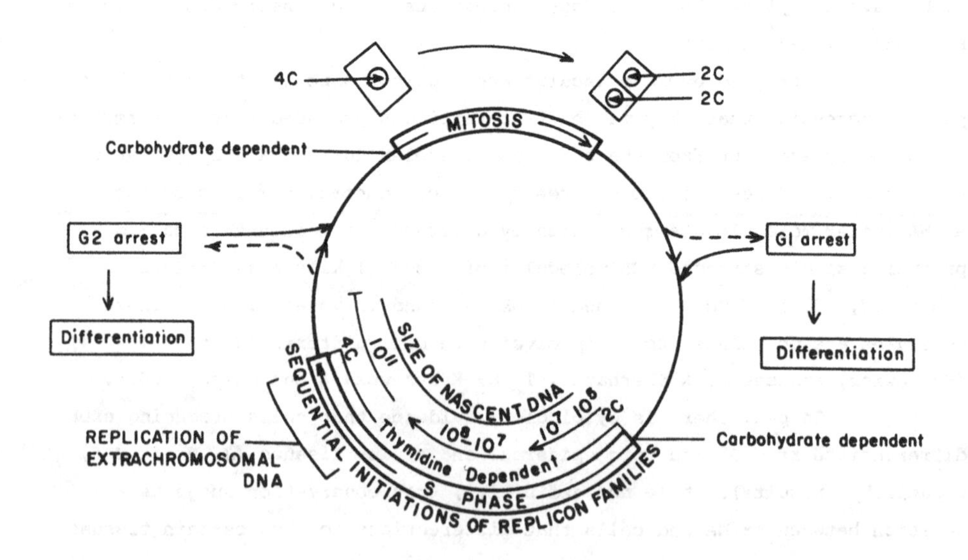

modified cell cycles (Dyer, 1976; see also Nagl, Pohl, Radler, this volume). The positions in G1 and G2 noted as carbohydrate-dependent are relative and they refer to the respective transitions from G1 to S and from G2 to M. Cells starved of either carbohydrate or phosphate fail to make these transitions. Those deprived of nitrate, however, fail only at G1 to S transition (Gould, Everett, Wang & Street, 1981) indicating that the controls that operate in G1 differ from those that operate in G2. This evidence and that of the temporal order of chromosomal DNA replication argues favourably for the view that the cell cycle is controlled by genes acting in sequence whose time of expression is determined by mitosis and the amount of nuclear DNA (2*C* vs 4*C*) in the cell.

ACKNOWLEDGEMENT

This research was supported by the U.S. Department of Energy.

REFERENCES

Beach, D., Durkacz, B. & Nurse, P. (1983). Functionally homologous cell cycle control genes in budding and fission yeast. *Nature*, 300, 706-9.

Bernier, G., Kinet, J.-M. & Bronchart, R. (1967). Cellular events at the meristem during floral induction in *Sinapis alba* L. *Physiologie Vegetale*, 5, 311-24.

Buchowicz, J., Kraszewska, E. & Eberhardt, J. (1978). Characterization of the early synthesized DNA in germinating *Triticum aestivum* embryos. *Phytochemistry*, 17, 1481-4.

Cairns, J. (1966). Autoradiography of HeLa cell DNA. *Journal of Molecular Biology*, 15, 372-3.

Dyer, A.F. (1976). Modification and errors of mitotic cell division in relation to differentiation. In *Cell Division in Higher Plants*, ed. M.M. Yeoman, pp. 199-249. London & New York: Academic Press.

Evans, L.S., Almheida, M.S., Lynn, D.G. & Nakaniski, K. (1979). Chemical characterization of a hormone that promotes cell arrest in G2 in complex tissue. *Science*, 203, 1122-3.

Evans, L.S. & Tramontano, W.A. (1981). Is trigonelline a plant hormone? *American Journal of Botany*, 68, 1182-9.

Evans, L.S. & Tramontano, W.A. (1984). Trigonelline and promotion of cell arrest in G2 of various legumes. *Phytochemistry*, in press.

Evans, L.S. & Van't Hof, J. (1973). Cell arrest in G2 in root meristems. *Experimental Cell Research*, 82, 471-3.

Evans, L.S. & Van't Hof, J. (1974a). Is the nuclear DNA content of mature root cells prescribed in the root meristem? *American Journal of Botany*, 61, 1104-11.

Evans, L.S. & Van't Hof, J. (1974b). Promotion of cell arrest in G2 in root and shoot meristems in *Pisum* by a factor from the cotyledons. *Experimental Cell Research*, 87, 259-64.

Francis, D. (1981a). Effects of red and far-red light on cell division in the shoot apex of *Silene coeli-rosa* L. *Protoplasma*, 107, 285-99.

Francis, D. (1981b). A rapid accumulation of cells in G2 in the shoot apex of Silene coeli-rosa during the first day of floral induction. Annals of Botany, 48, 391-4.

Francis, D. & Lyndon, R.F. (1978). Early effects of floral induction on cell division in the shoot apex of Silene. Planta, 139, 273-9.

Funderud, S., Andreassen, R. & Haugli, F. (1978). Size distribution and maturation of newly replicated DNA through the S and G2 phases of Physarum polycephalum. Cell, 15, 1519-6.

Gelfant, S. (1962). Initiation of mitosis in relation to the cell division cycle. Experimental Cell Research, 26, 395-403.

Gelfant, S. (1963). Patterns of epidermal cell division. I. Genetic behaviour of the G1 cell population. Experimental Cell Research, 32, 521-8.

Gelfant, S. (1966). Patterns of cell division: The demonstration of discrete cell populations. In Methods of Cell Physiology, vol. 2, ed. D.M. Prescott, pp.359-95. New York: Academic Press.

Gould, A.R., Everett, N.P., Wang, T.L. & Street, H.E. (1981). Studies on the control of the cell cycle in cultured plant cells. I. Effect of nutrient limitation and nutrient starvation. Protoplasma, 106, 1-13.

Hand, R. (1975). Regulation of DNA replication on subchromosomal units of mammalian cells. Journal of Cell Biology, 64, 89-97.

Harland, J., Jackson, J.F. & Yeoman, M.M. (1973). Changes in some enzymes involved in DNA biosynthesis following induction of division in cultured plant cells. Journal of Cell Science, 13, 121-38.

Howard, A. & Pelc, S.R. (1953). Synthesis of deoxyribonucleic acid in normal and irradiated cells and its relation to chromosome breakage. Heredity (Suppl.), 6, 216-73.

Huberman, A. & Riggs, A.D. (1968). On the mechanisms of DNA replication in mammalian chromosomes. Journal of Molecular Biology, 32, 327-41.

Jacqmard, A. & Miksche, J.P. (1971). Cell population and quantitative changes of DNA in the shoot apex of Sinapis alba during floral induction. Botanical Gazette, 132, 364-7.

Jacqmard, A., Raju, M.V.S., Kinet, J.M. & Bernier, G. (1976). The early action of the floral stimulus on mitotic activity and DNA synthesis in the apical meristem of Xanthium strumarium. American Journal of Botany, 63, 166-74.

Kowalski, J. & Cheevers, U.P. (1976). Synthesis of high molecular weight DNA strands during S phase. Journal of Molecular Biology, 104, 603-15.

Kraszewska, E. & Buchowicz, J. (1983). Uptake and binding of cytoplasmic DNA by wheat embryo cell nuclei. Molecular Biology Reports, 9, 175-8.

Krimer, D.B. & Van't Hof, J. (1983). Extrachromosomal DNA of pea (Pisum sativum) root-tip cells replicates by strand displacement. Proceedings of the National Academy of Sciences of the U.S.A., 80, 1933-7.

Mathews, C.K. & Sinha, N.K. (1982). Are DNA precursors concentrated at replicon sites? Proceedings of the National Academy of Sciences of the U.S.A., 79, 302-6.

Rost, T.L. (1977). Response of the plant cell cycle to stress. In Mechanisms and Control of Cell Division, ed. T.L. Rost & E.M. Gifford, Jr., pp.111-43. Stroudsburg Pa: Dowden, Hutchinson & Ross, Inc.

Schvartzman, J.B., Chenet, B., Bjerknes, C.A. & Van't Hof, J. (1981). Nascent replicons are synchronously joined at the end of S phase or during G2 phase in peas. Biochimica et Biophysica Acta 653, 185-92.

Schvartzman, J.B., Krimer, D.B. & Van't Hof, J. (1984). The effects of different thymidine concentrations on DNA replication in pea-root cells synchronized by a protracted 5-fluorodeoxyuridine treatment. Experimental Cell Research, 150, 379-89.
Skoog, K.L., Nordenskjold, B.A. & Bjursell, K.G. (1973). Deoxyribonucleoside triphosphate pools and DNA synthesis in synchronized hamster cells. European Journal of Biochemistry, 33, 428-32.
Taylor, J.H. (1963). DNA synthesis in relation to chromosomal reproduction and the reunion of breaks. Journal of Cellular and Comparative Physiology, 62, 73-86.
Van't Hof, J. (1965). Relationships between mitotic cycle duration, S period duration, and the average rate of DNA synthesis in the root meristem cells of several plants. Experimental Cell Research, 39, 48-54.
Van't Hof, J. (1973). The regulation of cell division in higher plants. Brookhaven Symposia in Biology, 25, 152-65.
Van't Hof, J. (1974a). The duration of chromosomal DNA synthesis, of the mitotic cycle, and of meiosis in higher plants. In Handbook of Genetics, ed. R.C.King, pp.363-77. New York: Plenum.
Van't Hof, J. (1974b). Control of the cell cycle in higher plants. In Cell Cycle Controls, ed. G.M. Padilla, I.L. Cameron & A. Zimmerman, pp. 76-86, New York: Academic Press.
Van't Hof, J. & Bjerknes, C.A. (1977). 18 μm replication units of chromosomal DNA fibers of differentiated cells of pea (Pisum sativum) Chromosoma, 64, 287-94.
Van't Hof, J. & Bjerknes, C.A. (1981). Similar replicon properties of higher plant cells with different S periods and genome sizes. Experimental Cell Research, 136, 461-5.
Van't Hof, J. & Bjerknes, C.A. (1982). Cells of pea (Pisum sativum) that differentiate from G2 phase have extrachromosomal DNA. Molecular and Cellular Biology, 2, 339-45.
Van't Hof, J., Bjerknes, C.A. & Clinton, J.H. (1978). Replicon properties of chromosomal DNA fibers and the duration of DNA synthesis of sunflower root-tip meristem cells at different temperatures. Chromosoma, 66, 161-71.
Van't Hof, J., Bjerknes, C.A. & Delihas, N.C. (1983). Excision and replication of extrachromosomal DNA of pea (Pisum sativum). Molecular and Cell Biology, 3, 172-81.
Van't Hof, J. & Kovacs, C.J. (1972). Mitotic cycle regulation in the meristem of cultured roots: The principal control point hypothesis. In The Dynamics of Meristem Cell Populations, ed. M.W. Miller & C.C. Keuhnert, pp.15-32. New York: Plenum.
Van't Hof, J., Kuniyuki, A. & Bjerknes, C.A. (1978). The size and number of replicon families of chromosomal DNA of Arabidopsis thaliana. Chromosoma, 68, 269-85.
Walters, R.A., Tobey, R.A. & Hilderbrand, C.E. (1976). Chain elongation and joining of DNA synthesized during hydroxyurea treatment of Chinese hamster cells. Biochimica et Biophysica Acta, 447, 36-44.
Walters, R.A., Tobey, R.A. & Ratliff, R.L. (1973). Cell-cycle-dependent variations of deoxyribonucleoside triphosphate pools in Chinese hamster. Biochimica et Biophysica Acta, 319, 336-47.

THE ORGANIZATION OF REPLICONS

J.H. Waterborg and S. Shall

INTRODUCTION

This paper is a collection of ideas related to the concept of a unit of DNA replication, the replicon. The various levels of organization of chromatin and nucleus together ensure the correct regulation, both temporal and spatial, of the process of replication of all the DNA molecules, and their equal distribution to the two daughter cells during mitotic division.

The experiments with which we will illustrate this discussion are mainly taken from studies with the lower eukaryote *Physarum polycephalum*. This organism has a very high natural mitotic synchrony which allows many experimental approaches to the regulation of DNA replication that are impossible in other cell systems.

MOLECULAR STRUCTURE OF A REPLICON

Definition of a replicon

A replicon is a unit of replication of DNA. It may be detected by DNA fibre autoradiography. Its length is the distance between two neighbouring sites of initiation of DNA synthesis on the same DNA molecule (Bryant, 1982; Francis, Kidd & Bennett, this volume; Van't Hof, this volume). The length may also be measured by electron microscopy of replicating DNA molecules as the centre-to-centre distance between replication eyes (Evans, 1982). In sedimentation studies of newly synthesized DNA it is recognized as a relatively stable intermediate (see Van't Hof, this volume).

The size of replicons is variable between organisms (Francis, Kidd & Bennett, this volume). Small replicons are seen in *Physarum* with a size of 30-50 kbp (10-15 μm) of DNA (Funderud, Andreassen & Haugli, 1978a, 1978b, 1979). Large replicons with up to 600 kbp (200 μm) of DNA occur in *Triturus cristatus-carnifex* (Buongiorno-Nardelli, Micheli, Carri & Marriley, 1982). During early development, replicon sizes may increase as in *Drosophila* where the average replicon size increases from 10 kbp (3 μm) to 40 kbp (14 μm) (Blumenthall, Kriegstein & Hogness, 1974). Average values for replicon size are generally between 50 and 100 kbp (15-50 μm) (Edenberg & Huberman, 1975). The knowledge of replicon size and the size of the genome of an organism allows one to calculate the number of replicons in a cell. As an example, such calculations are shown for the acellular myxomycete *Physarum polycephalum* in Table 1. One genome of DNA sequences (haploid G1 phase; *C* value) contains approximately 5600 replicons distributed over approximately 40 DNA molecules or chromosomes. For other examples see Bryant (1982) and Francis, Kidd and Bennett (this volume).

Initiation of DNA replication in S-phase

Each DNA molecule in a cell is duplicated in its entirety once, and not more than once, in each cell cycle. This suggests a very tight control over the initiation of replicon duplication. In one way or another the cell marks the site of initiation of DNA synthesis after DNA synthesis has begun, to prevent re-initiation of the same replicon within the same cell cycle (Wanka, 1984). The cell presumably also prevents initiation of 'late' replicons when the replication fork from an 'early' replicon has passed and already has duplicated the DNA sequence. However, in a normal cell cycle no replicon may be left unduplicated (but see Nagl, Pohl & Radler, this volume). Although the precise mechanisms are not yet clear, a number of recent observations in *Physarum* have given us some idea as to how these events are accomplished.

Funderud *et al*. (1978a) and Haugli, Andreassen & Funderud (1982), have described observations which suggest a model that they have called 'master initiation' (Fig. 1, stage 2). *Physarum* macroplasmodia are giant, multinuclear cells with a natural mitotic synchrony. DNA synthesis starts at about eight minutes after metaphase (Beach, Piper and Shall, 1980a). Bromodeoxyuridine (BUdR) was incorporated for a very short period at the very start of S-phase. The remaining new DNA was labelled with radioactive deoxyadenosine (AdR) and the DNA was isolated and sized by alkaline sucrose gradient centrifugation. Exposure of the BUdR to ultra-violet light fragmented the full-sized DNA to replicon size (30-65 kbp). In addition to early replicated DNA, late replicated DNA was also fragmented to replicon size. They concluded that apparently at the beginning of S-phase, short patches of BUdR-containing DNA had been synthesized at all the origins (Fig. 1, stage 2). After this initiation, only the early replicons had continued DNA synthesis

Table 1A Levels of Chromatin Organisation in a Nucleus of *Physarum polycephalum*

	Nucleus (4C)	Genome (1C)	DNA Molecule	Loop/ Replicon	30 nm Turn	Nucleosome	Base Pair of DNA
Chromosomes	80	40	1	-	-	-	-
Molecules	160	40	1	-	-	-	-
DNA Loops	22 500	5 600	175	1	-	-	-
Nucleosomes	6.4×10^{6}	1.6×10^{6}	40×10^{3}	230	6.5	1	-
Base Pairs (bp)	1.1×10^{9}	280×10^{6}	7.0×10^{6}	$\underline{40 \times 10^{3}}$	1.1×10^{3}	172	1
30 nm Fibre (length)	11.0 mm	2.8 mm	68 µm	390 nm	<u>11 nm</u>	-	-
11 nm Fibre (length)	70.0 mm	18.0 mm	440 µm	2.5 µm	71 nm	<u>11 nm</u>	-
DNA Helix (length)	370 mm	90 mm	2.3 mm	13 m	365 nm	57 nm	<u>0.33 nm</u>
Linear Condensation of DNA	(72 000 X)	(18 000 X)	2100 X	190 X	33 X	5 X	1 X
DNA (g)	1.1×10^{-12}	$\underline{280 \times 10^{-15}}$	7.0×10^{-15}	40×10^{-18}	1.1×10^{-18}	170×10^{-21}	1×10^{-21}
DNA (D)	740×10^{9}	185×10^{9}	4.6×10^{9}	26×10^{6}	725×10^{3}	115×10^{3}	660
References	-	A,B	A	B,C	D,E	D,E	E

4C DNA content: diploid nucleus at G2-phase or metaphase.
1C DNA content: genome complexity, haploid nucleus at G1-phase.

Nucleus	Structure and Dimensions	References	References
Nucleus	Sphere with a diameter of <u>5000 nm</u>	Association of approximately 80 chromosomes with nuclear matrix and envelope.	A
Chromosome	Rod, length <u>1100 nm</u>, diameter <u>450 nm</u>.	Loops of 30 nm fiber chromatin on scaffold. Condensed (see dimensions) in metaphase. No chromatids visible in metaphase chromasomes.	A
DNA loop	DNA loop, diameter <u>70 nm</u>.	Attached loop of 30 nm chromatin fiber. Contracted to super-solenoid in metaphase (see dimension). Possibly identical to replicon and/or transcriptional domain.	F,G
30 nm Fiber	Solenoid, diameter <u>30 nm</u>, pitch <u>11 nm</u>.	11 nm fiber with 6 to 7 nucleosomes per turn. 'solenoid' could contain superbeads.	D,G
11 nm Fiber	Linear chain, diameter <u>11 nm</u>.	Chain of nucleosomes, 'beads-on-a-string'. Potentially active chromatin?	G
Nucleosome	Disk, dimensions <u>11 x 11 x 6 nm</u>.	Approximately 172 base pairs of DNA + 8 core histones (H2A + H2B + H3 + H4) + histone H1.	E,G
DNA (B-DNA)	Alpha helix, diameter <u>2 nm</u>, pitch <u>3.4 nm</u>.	Approximately 10.5 base pairs of anti-parallel DNA per turn of helix.	E,G

Underlined data were used for the calculations and are from references cited.

REFERENCES
A = Mohberg (1977, 1982); B = Holt (1980); C = Evans (1982); D = Matthews & Waterborg (1984);
E = Matthews & Bradbury (1982); F = Adolph & Kreisman (1983); G = Butler (1984)

bidirectionally, while all the other replicons were held in an initiated state (Fig. 1, stage 3) until a second signal allowed these to continue chain elongation (Willie & Kauffman, 1975).

This explanation of their observations suggests that within a very short interval all origins are initiated. It is thought that this initiation changes the structure of the DNA at these sites in such a way that no re-initiation can occur within the same cell cycle (Wanka, 1984). As will be discussed below, the replicon origins or sequences very near them,

Fig. 1. A schematic model of DNA replication in the Physarum cell cycle. Replication of three clusters of replicons (A-C) containing replication origins (P-X). Stages 1-8 exist at the time indicated, relative to the beginning of S-phase, immediately after mitosis. (1) t = 0 min: DNA prior to the start of S-phase and in G2-phase. (2) t = 1 min: Master initiation of all active origins at start of S-phase, start of DNA synthesis in cluster B. Origin W is an inactive origin due to matrix detachment. (3) t = 20 min: Okazaki synthesis and ligation in cluster B, start of synthesis in cluster C. (4) t = 40 min: termination in cluster B, Okazaki synthesis in cluster C. (5) t = 60 min: ligation in cluster B, termination in cluster C except at origin W, termination of replicon Q at initiated origin P, start of synthesis in cluster A. (6) t = 90 min: ligation in cluster C, termination at origin W, synthesis in cluster A. (7) t = 120 min: ligation of new, cluster sized DNA. (8) t = 8 hours: mitotic segregation of chromosomes.

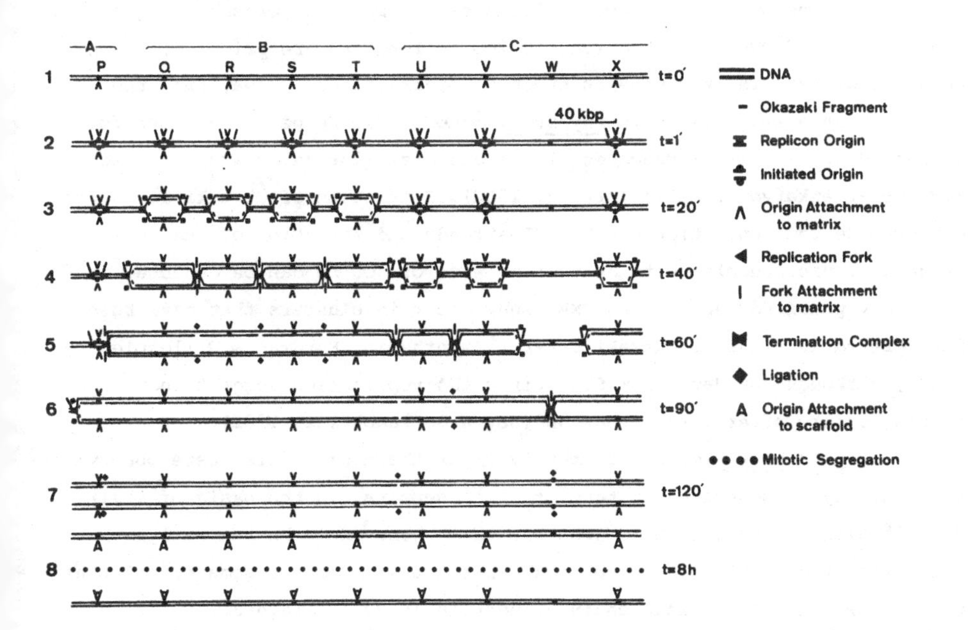

may be attached to the nuclear matrix. The interaction between origin and matrix could change irreversibly upon initiation, for instance by formation of the presumed 'eye' and doubling of the attachment structure. The regulation of the initiation of replicons may also require soluble proteins that are made during only a very small time interval of the cell cycle.

Physarum is the only organism so far which suggests 'master initiation' of replicon origins. The conventional view that replication of the origin is part of the complete replicon duplication and is therefore sequential, as replicon synthesis itself, in S-phase (see below), may still be valid for other organisms although no experimental support for this view is known.

DNA synthesis and termination in S-phase

DNA synthesis is, at least partially, a discontinuous process. DNA is synthesized by DNA polymerase from nucleoside triphosphates using the opposite strand of the DNA as template. In many cells the 'leading' strand synthesis is continuous (Kornberg, 1980). The 'lagging' strand can only be synthesized discontinuously, since DNA polymerase can only add to a 3' terminus. It seems however that in *Physarum*, *both* strands are synthesized discontinuously (Holt, 1980; Evans, 1982); this may also be so in higher plants (Bryant, 1982).

The smallest discrete pieces of new DNA are 'Okazaki fragments'. These pieces of DNA are approximately 200 bp long and are primed by a short sequence of RNA. Their synthesis proceeds bidirectionally away from the origin at an average rate, in *Physarum*, of 600-1200 bp per minute per fork (Funderud *et al.*, 1979; Beach *et al.*, 1980a); similar rates are observed in many other eukaryotic cells (Bryant, 1982). Initiation of Okazaki fragments may precede the separation of the DNA strands and formation of the fork structure. Microbubbles with an average size of 500 bp can be detected in carefully prepared replicating DNA. They occur in clusters that have been interpreted as sites of Okazaki pre-fork synthesis (Hardman & Gillespie, 1980), although the data from SV40 virus DNA replication suggest that priming occurs after fork formation (Hay & De Pamphilis, 1982).

In *Physarum*, approximately 2% of the microbubbles were observed as single eyes, separated by tens of kilobasepairs. On the basis of their distribution in the DNA it is possible that these could be the early-initiated origins of 'late' replicons that are waiting for some signal that will allow them to continue chain elongation (Fig. 1, stage 3).

Termination of DNA synthesis in a replicon apparently occurs randomly. Specific termination sites or sequences have never been observed. Termination will thus generally occur approximately halfway between adjacent origins, when replication forks meet (Fig. 1, stage 4). The molecular process of termination is unknown. Presumably the replication complexes dissociate, a repair-like process fills the gap and ligase joins the replicon-sized DNA (Fig 1, stage 5). The synthesis of DNA from Okazaki pieces to the replicon size of approximately 40 kbp in *Physarum* is complete in 40 to 50 minutes for those replicons that are initiated at the beginning of S-phase. The increase in the size of this newly synthesized DNA then stops for about 30 minutes. It then increases stepwise to the size of a replicon cluster: three or four replicons. After a further interval until nearly the end of S-phase, two hours after mitosis, it finally reaches full size DNA (Fig. 1, stage 7). Condensation of newly synthesized DNA to intermediate and full length fragments seems to be a rather slow process compared with the rate of chain elongation. The ligation of new DNA may continue into G2-phase (Funderud *et al.*, 1978a, 1978b, 1979); this has also been observed in higher plants (Van't Hof, this volume).

The temporal organization of S-phase

The order in which replicons are duplicated in S-phase, is fixed. This was shown in *Physarum* by radioactive labelling and density labelling of replicating DNA at identical times within the S-phase of two successive cell cycles. All the labelled DNA was double-labelled (Holt, 1980; Evans, 1982). Mixing experiments with nuclei and cytoplasm from different phases of the DNA replication process showed that cytoplasmic, cell-cycle-specific factors are essential for the correct progression of replication (Wille & Kauffman, 1975; Wille, 1977; Holt, 1980).

A fixed temporal order in S-phase appears to be a general phenomenon. In yeast, functional origins (autonomously replicating sequences) ars, are activated in a specific temporal order during the cell cycle (Chan & Tye, 1983). Sister chromosomes of diplochromosomes replicate synchronously in identical patterns (Lau, 1983). Defined temporal groups or 'families' of replicons can often be distinguished; for example there are two in *Arabidopsis thaliana* and 10 in *Physarum* (Muldoon, Evans, Nygaard & Evans, 1971; Van't Hof, Kuniyuki & Bjerknes, 1978). This temporal organization results in *Physarum* in the observation that in the first quarter of S-phase little or no repetitive DNA is replicated (Evans, 1982). Generally, euchromatin appears to be replicated earlier than heterochromatin, transcribed sequences before untranscribed ones and constitutive housekeeping genes before tissue specific genes (Bello, 1983; Holmquist & Goldman, 1983).

DNA synthesis outside of S-phase

Some DNA synthesis occurs outside of S-phase. The DNA coding for ribosomal RNA in *Physarum* exists in many copies of a non-chromosomal, linear palindromic DNA of about 60 kbp. They are replicated from a replication origin near the centre of the palindrome by bidirectional synthesis. The control over the origin of this extra-chromosomal replicon appears to be different from that of the chromosomal replicons. Replication occurs essentially all through the cell cycle. In each cycle some molecules are replicated once, some are replicated twice, others are not replicated at all, with the result that on average a constant ratio between chromosomal DNA and extra-chromosomal 'ribosomal' DNA is maintained. In this 'balance' the selection of 'ribosomal' DNA molecules to be replicated appears to be random (Vogt & Braun, 1977; Holt, 1980).

Isolation of origins

DNA synthesis begins at defined sites or 'origins' in prokaryotic chromosomes, such as those of *E. coli* and its viruses. Similarly, origins of replication have been defined in some eukaryote viruses such as SV40,polyoma, BK and adenoviruses, and in the mitochondrial DNA of animal cells. However, no origin of replication from a eukaryotic chromosome has thus far been unequivocally defined. It therefore remains unproven whether or not DNA synthesis is initiated at precisely defined nucleotide sequences.

Despite the considerable amount of work on replicons and replication of eukaryotic DNA, very little is known about the nature of eukaryotic chromosomal origins of replication, primarily due to the absence of an assay system which responds to *bona fide* origins of replication, a problem still not fully resolved. Work on eukaryotic origins of replication has therefore been largely confined to extra-chromosomal molecules. In these cases an organization similar to that in prokaryotic replicons has been found with specific sequences being required for the initiation of replication (Soeda, Arrand, Smolar & Griffin, 1979; Clayton, 1982).

The development of a transformation system for yeast appeared to offer a means of assaying eukaryotic origins of replication (Hinnen, Hicks & Fink, 1978; Beggs, 1978). This has permitted the isolation of such autonomously replicating sequences (ars) from the chromosomal DNA of *Saccharomyces cerevisiae* (Struhl, Stinchcomb, Scherer & Davis, 1979; Beach *et al.*, 1980b). The frequency of these sequences in the *S. cerevisiae* genome was found to be similar to the average spacing of replicons (Beach *et al.*, 1980b; Chan & Tye, 1980); they contained similar short AT-rich blocks within a larger required sequence as do isolated non-chromosomal origins (Broach *et al.*, 1982), and consequently, by analogy with the situation in prokaryotes, it was suggested that these special sequences may be eukaryotic origins of DNA replication. The ars assay in yeast also yielded active ars from other eukaryotic DNA, such as *Physarum* (Gorman, Dove & Warren, 1981), but not from *E. coli* DNA. Recent experiments (K. Maundrell & S. Shall, unpublished data) showed that out of a random selection of sequences from *Schizosaccharomyces pombe*, the expected number had ars activity in *S. pombe*. Testing all the sequences in the heterologous system of *S. cerevisiae* showed that less than half of the ars sequences in *S. pombe* were recognized as such by *S. cerevisiae*, while a certain number of the non-ars sequences from *S. pombe* showed ars activity in *S. cerevisiae*. This result suggests that a heterologous assay for ars activity may be of doubtful value for isolation of eukaryotic origin sequences.

Origins of DNA replication were isolated by us using a totally independent principle (Beach *et al.*, 1980a). The protocol relied on isolating that DNA which was synthesized in the first few minutes of S-phase in *Physarum*. This DNA was density labelled with BUdR, isolated and analyzed by equilibrium centrifugation in caesium chloride density gradients. The analysis suggested that gentle shearing was probably breaking the DNA molecule selectively at the replication fork, thus freeing double-stranded, newly replicated DNA from the flanking unreplicated part of the molecule. We have not rigorously excluded the possibility of branch migration, but our evidence is that we obtained heavy-light hybrid molecules and not heavy-heavy ones expected as a product of branch migration. Sheared, newly initiated DNA thus consists of molecules of two size classes: bulk DNA with a molecular weight greater than 45 kbp, and newly replicated DNA whose size increases by 2.4 kbp per minutes from the beginning of S-phase, as expected for known fork rates in *Physarum*.

Our data further suggested that at the start of DNA replication there is a single initiation event simultaneously at approximately 3000 'origins' of replication (see Table 1), and that little or no further initiation occurs for at least 15 minutes.

A method for the purification, isolation and molecular cloning of this DNA has been developed. Subsequently, we have established that these newly initiated DNA fragments can be more readily isolated and cloned by tritium pulse labelling followed by controlled shear and isolation on a sucrose gradient. After 2 minutes of S-phase this procedure yields DNA fragments with an average size of about 3 kbp.

MACROSCOPIC STRUCTURAL FEATURES OF REPLICONS

The requirement for a structural organization of the nuclear DNA

In eukaryotic cells very large amounts of DNA are packaged within the very small volume of a nucleus. For an example, see Table 1. Once during every cell cycle the amount of DNA is doubled. The new DNA molecules are segregated equally over two daughter nuclei during mitosis. Apparently a minimal number of DNA strands are found at that point in the cell cycle to be entangled despite the fact that DNA replication requires the complete unwinding of all the DNA molecules down to the last helical turn. These highly complex processes that are interdependent, and also dependent on the transcriptional regulation of the genome lead one to expect a rigorous but flexible pattern of organization (Wanka, 1984).

Chromatin organization of DNA

Much is already known about the lower levels of organization of the DNA into chromatin. (For some recent reviews, see Matthews & Bradbury, 1982; Nagl, 1982; Butler, 1984; Matthews & Waterborg, 1984). In general, approximately 200 base pairs of DNA are associated with an octamer of core histones and one molecule of H1 to form nucleosomes. Most nucleosomes are organized in a fibre with a diameter of approximately 30 nm. This fibre is formed by the i eractions between the H1 molecules of the nucleosomes and by the interaction of the amino terminal regions of some of the core histones with the DNA of neighbouring nucleosomes. The packing of DNA into a 30 nm fibre gives a linear contraction of about 40 (Table 1).

To allow replication of chromatin, the 30 nm fibre is dis-assembled into a 'beads-on-a-string' structure of nucleosomes on DNA. This causes an increase in the sensitivity of the DNA to nuclease digestion. Nucleosomes appear to remain associated with the DNA all through the process of replication, except in the immediate location of the replication fork. Re-packaging of the DNA with pre-existing and newly formed nucleosomes is followed by a slow process of maturation. In the 30 minutes following DNA synthesis, the spacing of the nucleosomes, the 30 nm chromatin fibre and the low nuclease sensitivity of the chromatin are all restored (Matthews & Waterborg, 1984). This process of unfolding and reformation of chromatin causes a variation in the DNase I sensitivity of the chromatin through the cell cycle that can be observed in *Physarum* (Jalouzot *et al.*, 1980).

Nuclear matrix and chromosome scaffold

Folding of the 30 nm fibre, the highest reproducibly recognizable chromatin structure, requires the interaction between DNA and other proteins such as those of the chromosomal 'scaffold' (Barrack & Coffey, 1982; Nagl, 1982; Butler, 1984; Wanka, 1984). This is a limited set of proteins, constituting approximately 3-4% of the total chromosomal protein, that are isolated from metaphase chromosomes after removal of histones and DNA by high salt and nuclease digestion. Together they form a three dimensional network that has the shape of the metaphase chromosome before the chromatin was removed. The structural integrity of this network depends on metallo-protein interactions involving Cu^{2+} and possibly Ca^{2+} ions and/or disulphide bonds between a subset of the proteins. The network has contractile properties that may require phosphorylation of scaffold proteins during mitosis (Barrack & Coffey, 1982). This contraction, in combination with contraction of the chromatin by phosphorylation of H1 (Matthews & Bradbury, 1982), yields the highly condensed metaphase chromosomes in which the linear compaction ratio of the DNA is increased to several thousand fold (Table 1). In mammalian cells this compaction ratio can reach, at metaphase, values of nearly 10,000 when 0.26 pg of DNA in 2 molecules, the average value for DNA in a human metaphase chromosome, i.e. 86 mm of DNA, is compacted into a 10 micron-long chromosome.

After mitosis a nuclear envelope with its associated pore complex lamina reforms around the two separated masses of telophase chromosomes. The chromatin in the new nuclei decondenses. Isolation of the 'scaffold' at this stage of the cell cycle, by high salt and nuclease treatment, yields a nuclear ghost or nuclear matrix which, just as the scaffold does during mitosis, retains the shape and structure of the nucleus. However, the protein composition is more complex, containing as much as 10-15% of the total nuclear proteins. It consists of the scaffold proteins, the proteins that form the pore and lamina structure at the nuclear envelope, and an abundance of minor structural, and probably also regulatory, proteins that are involved in all the processes that occur localized on the nuclear matrix structure (Barrack & Coffey, 1982). In Physarum with its closed mitosis, the composition of the nuclear matrix proteins in interphase is virtually indistinguishable from that in mitosis (Wanka, 1984). Although the continuity between chromosomal scaffold and nuclear matrix has not been proven directly, due to the difficulties in isolating the intermediate stages at the beginning and end of mitosis, observations of prematurely condensed chromosomes, made by fusion of mitotic and interphase-cells in G1, S or G2 phase, supports this idea (Hanks et al., 1983; Wanka, 1984).

DNA association with the nuclear matrix

DNA is bound, as chromatin, to the chromosome scaffold and nuclear matrix structures in long loops of several tens of kilobasepairs. The size of these loops is identical on both structures, further supporting the idea of identity between scaffold and matrix (Butler, 1984; Wanka, 1984). Scanning electron microscopy shows these loops in human metaphase chromosomes as short stubby, radially orientated rods with diameters of approximately 70 nm and formed from a supertwisted 30 nm chromatin fibre, as judged after partial decondensation of the chromatin loop (Adolph & Kreisman, 1983; Hanks et al., 1983). These loops may be organized in groups that form the bands that can be visualized by staining of the metaphase chromosomes (Butler, 1984).

Removal of the chromatin proteins by high salt treatment allows the visualization of the DNA in the loop itself. Observation of an interphase nuclear matrix by fluorescence microscopy in the presence of ethidium bromide shows quite clearly a halo of DNA round the nuclear ghost structure. The diameter of this halo is dependent on the ethidium bromide concentration (Vogelstein, Pardoll & Coffey, 1980). This shows that all the DNA loops are separate supercoiled domains. Digestion of the loops by nucleases results transiently in the maximal size of the halo. This allows a determination of the loop size. The loops are generally about 50 to 100 kbp long but this value varies among different organisms (Buongiorno-Nardelli et al., 1982; Hyde, 1982; Butler, 1984).

Preparation of a nuclear matrix retaining all its DNA (Wanka, 1984) depends on the absence of shear and of endogenous nuclease activity during the high salt treatment of the nuclei. A standard protocol treats isolated Physarum nuclei with buffered 2 M NaCl followed by the isolation of the residual, rapidly sedimenting structure by sedimentation on top of a cushion or as a pellet (Aelen, Opstelten & Wanka, 1983). Such a preparation gives maximally 90% of the DNA retained in the isolated matrix, and routinely 60% or more. Inclusion of polyethyleneglycol in such a protocol (Table 2) diminishes the shear forces on the DNA during the removal of the histones by 2 M NaCl, abolishes the shear forces on the DNA loops during sedimentation of the matrix surrounded by its halo, and suppresses endogenous nuclease activity. Routinely, more than 98% of the DNA is retained on such a matrix and the dependence of the halo size of a matrix, reswollen in a low ionic strength buffer, upon the concentration of ethidium bromide indicates the absence of nicking in the supercoiled DNA loops.

Preparation of nuclear matrix structures that retain all their DNA, allows the subsequent analysis of the DNA attached at or very close to the matrix. The loop DNA can be removed by nonspecific nucleases like DNase I or micrococcal nuclease or by restriction enzymes. The residual DNA is then isolated by sedimentation of the matrix (Table 2). In many organisms the residual DNA, 1% or less, does not differ from total, bulk DNA as judged by solution hybridization, although frequently, enrichment of satellite or 'ribosomal' DNA sequences on the matrix is observed (Wanka, 1984). Such enrichments could be real, i.e. these sequences could be consistently located in the DNA loops at or near the matrix, or they may depend on the sequence specificity of the nuclease used, since such satellites have only a limited sequence complexity.

So far hybridization studies have not been able to detect sequences that are common among the attachment sites of DNA loops to the matrix. This means that such a sequence does not exist or that it is too small to be detected by hybridization. Direct sequence determination of matrix DNA isolated from the G1/S boundary of mouse 3T3 cells has detected short blocks of repetitive sequences that were highly enriched on the matrix. These sequences contained features that may suggest the presence of replicon origins: AT-rich blocks and GC/AT switches with purine bias occur as in functional ars sequences; viral large tumour antigen recognition sequences are present, and the sequences display the possibility for secondary DNA structure (Goldberg, Collier & Cassel, 1983).

These structures, or short blocks of primary DNA sequence, may direct the binding of those DNA sites to specific proteins. In HeLa cells proteins have been detected that bind to DNA very tightly at sites spaced apart by approximately loop size distances (Barrack & Coffey, 1982; Bodnar _et al._, 1983). Many other proteins in the nuclear matrix have DNA binding

Table 2 Preparation of nuclear matrix DNA of Physarum

MACROPLASMODIUM

Homogenize in 0.25M sucrose 5mM $MgCl_2$ (or 10mM $CaCl_2$) - 10 mM Tris-0.1% Triton X100-pH 7.2

NUCLEI

+ homogenization buffer to 200,000,000 nuclei/ml.
+ 2 volumes 3M NaCl-15% PEG 6000-5mM EDTA-50mM Tris-pH 7.4
15 min on ice.
spin 1 min 3000 xg.

NUCLEAR MATRIX (contracted).

+ vortex in 0.5M NaCl-10% PEG 6000-5mM EDTA-20mM Tris-pH 7.4, at 2,000,000 per microliter.
+ 4 volumes 13.75mM $MgCl_2$-20mM Tris-pH 7.4
15 min on ice.

NUCLEAR MATRIX (swollen)

Digest 2 h 37 C with 0.5 Units Mbo 1 per µg DNA.
Optional: add 1 volume 4 M NaCl.
Spin 20 min 10,000xg through sucrose cushion

NUCLEAR MATRIX (digested)

Proteinase K digestion in 1% Sarkosyl.
Phenol extraction
Electrophoretic removal of slime
RNase A digestion

NUCLEAR MATRIX DNA

properties. Reconstitution of urea-dissociated matrix structures in the presence of bulk or matrix-associated DNA resulted in rapidly sedimenting structures that had bound DNA. However, no sequence specificity was apparent by solution hybridization (Mullenders, Eygenstein, Broen & Wanka, 1982).

In addition to DNA attachment to the internal nuclear matrix, DNA may be associated with the nuclear envelope, its pores and lamina. Such associations, e.g. of telomeric DNA sequences with the pore complex lamina (Franke, Scheer, Krohne & Janasch, 1981; Wanka, 1984), may play a role in recombination during meiotic prophase by allowing the pairing of homologous chromosomes. This association must be transient as the pore complex lamina dissociates during an open mitosis (Wanka, 1984).

Localization of DNA replication on the nuclear matrix

Radioactive labelling of newly synthesized DNA followed by autoradiography of nuclear halos or by sedimentation analysis for enrichment of label at the matrix during progressive detachment of loop DNA, has proved that DNA replication occurs in association with the nuclear matrix. The recent visualization of matrix DNA from S-phase HeLa cells, showed directly the enrichment for replication fork structures, as expected (Valenzuela, Mueller & Dasgupta, 1983).

A short pulse treatment with radioactive nucleotides during DNA synthesis, shows label exclusively localized above the nuclear ghost structure. This shows that all the DNA synthesized within the short period of the pulse, which can be given at any point within S-phase, is closely associated with the nuclear matrix. Very short pulses have suggested that replication occurs all through the nuclear volume and not exclusively near the pore complex lamina or the nuclear envelope, as some results have previously been interpreted (Wanka, 1984). Longer labelling gives a progressive appearance of label above the halo of DNA until by approximately 30 minutes equilibrium has been reached. By that time replicon size DNA has been synthesized (Barrack & Coffey, 1982).

Progressive detachment of loop DNA by DNase I digestion of nuclear matrix structures in cells, labelled continuously with ^{14}C-thymidine and pulse labelled with ^{3}H-thymidine at any point within S-phase, shows a gradual enrichment for newly replicated DNA at the nuclear matrix by the increasing ^{3}H/^{14}C-ratio (Dijkwel, Mullenders & Wanka, 1979; Wanka, 1984).

Pulse labelling at specific times in S-phase in Physarum, followed by a chase into G2-phase, allows the localization of a small patch of newly replicated DNA relative to the matrix. DNA labelled at the start of S-phase displays a continuous increase in the $^{3}H/^{14}C$ ratio upon nuclease digestion. It must be located right at the nuclear matrix, and it must remain there in G2-phase (see next section). Pulse labelling at the time between initiation and termination of the 'early' replicons (Fig. 1) results, after a chase, in a $^{3}H/^{14}C$ ratio that initially increases upon nuclease digestion but subsequently decreases again. Such a patch must be located halfway between matrix and the end of the DNA loop. DNA labelled at the end of S-phase, i.e. near the time of termination of the last replicons, displays a continuous decrease in the ratio. The termination sequences must be near the end of the DNA loops (Aelen et al., 1983; Wanka, 1984).

In regenerating rat liver, but not in normal liver, a replication complex that contains DNA polymerase -α, can be isolated in association with the nuclear matrix. Such complexes may contain several replication forks and may be the explanation that replicon synthesis is observed in small clusters, e.g. three to four replicons per cluster in Physarum (Funderud et al., 1979; Barrack & Coffey, 1982; Smith & Berezney, 1983). The replication complex is not a permanent structure in the cell cycle. It forms just prior to the start of DNA replication (Fig. 1, stage 1)(Smith & Berezney, 1983), possibly just behind the replication fork (Dijkwel et al., 1979). Observation of prematurely condensed chromosomes made with S-phase cells has suggested that complex formation may change the interaction of the DNA with the scaffold proteins (Hanks et al., 1983).

Viral and plasmid replication may also be localized on the nuclear matrix. Replication of SV40 occurs in association with the nuclear matrix (Barrack & Coffey, 1982). The presence of a yeast ars, a functional origin, in a plasmid DNA causes the intra-nuclear replication of such a plasmid (Kingsman, Clarke, Mortimer & Carbon, 1979).

Permanent attachment of DNA to the nuclear matrix through the replicon origins

The loop size of DNA, measured by the size of a DNA halo around a nuclear matrix, is very clearly correlated with the size of replicons over the whole known range of replicon size from 20 kbp to 600 kbp. In Xenopus the size of the average replicon increases during embryonal development and the diameter of the halo increases in parallel (Buongiorno-Nardelli et al., 1982).

The size of the DNA halo remains unchanged throughout the cell cycle. This indicates that attachment, which is more permanent than that exhibited by replication complexes, must determine the constant size of the DNA loops (McCready et al., 1980). Measurements of the number of permanent attachment sites per replicon vary around the value of one per replicon. McCready et al., (1980) observe in HeLa cell nuclear matrices (cages) one site per four replicons; Vogelstein et al., (1980) calculate one site per replicon from determination of the size of supercoiled domains by restriction nuclease digestion. Buongiorno-Nardelli et al., (1982) interpret their data on halo size as two attachment sites per replicon (reviewed by Wanka (1984)). Clearly more experimental evidence in systems with low variation in replicon or loop size is required to substantiate the idea that DNA is permanently attached to the nuclear matrix at one site per replicon.

Pulse-chase labelling at the start of S-phase in Physarum (see above) suggests that the site of permanent DNA attachment to the matrix is identical to a replicon origin (Aelen et al., 1983). DNA labelled at the beginning of S-phase remains after synthesis associated with or located near the matrix, irrespective of the length of chase. This suggests that the origin is at or very near to the permanent attachment point of the DNA loop to the matrix (Fig. 1). This conclusion is valid for the origins of the first temporal set of replicons of the S-phase of Physarum. The experiment does not prove that all origins, even origins of the first temporal family of replicons, are attached (Wanka, 1984). However, the observation that at the beginning of S-phase all origins are initiated in Physarum in the process called 'master initiation', may suggest the correctness of the observation for all origins. Isolation of origins from early and late replicons followed by the analysis of their location in the DNA loops relative to the nuclear matrix, may resolve this uncertainty.

In a different model, permanent replication complexes are assumed to explain the permanent attachment of DNA loops to the matrix. DNA replication at the matrix starts at the DNA between two adjacent replication complexes that have met at termination of replicon synthesis in the previous cell cycle. The DNA between the complexes, the functional origin of the replicon, is duplicated and two loops of DNA are reeled through the complexes. After completion of replicon synthesis the origin is far away from the matrix at the top of the DNA loop. The initiation complex for the next S-phase is formed between two replication complexes when they meet at termination (Vogelstein et al., 1980; Barrack & Coffey, 1982). The observation by Smith and Berezney (1983) that the replication complex on the matrix is not stable but reforms before every S-phase, does not support this model. Neither can it be reconciled with the observed pulse-chase labelling of DNA relative to the nuclear matrix at the beginning or at the end of the S-phase in Physarum.

The observation of increased replicon length and increased halo size during embryonic development, for instance in Xenopus (Buongiorno-Nardelli et al., 1982), in combination with the hypothesis that DNA is attached to the matrix at or near the origins of the replicons, suggests that the interaction between origin and matrix may be required for initiation. Such a system, in combination with 'master initiation' of origins, would be fail-safe since failure of initiation of an origin still leads to the correct duplication of all the DNA. If an origin has become detached accidently or as programmed during development, or has failed to initiate at the beginning of S-phase for some other reason, the replication fork of the neighbouring replicon continues chain elongation past the inactive initiation sequence until it meets another replication fork. Termination follows then as normal (Fig. 1, origin W).

ACKNOWLEDGEMENTS

J.H.W. thanks the European Molecular Biology Organization for an EMBO fellowship.

REFERENCES

Adolph, K.W. & Kreisman, L.R. (1983). Surface structure of isolated metaphase chromosomes. Experimental Cell Research, 147, 155-66.

Aelen, J.M.A., Opstelten, R.J.G. & Wanka, F. (1983). Organization of DNA replication in Physarum polycephalum. Nucleic Acids Research, 11, 1181-95.

Barrack, E.R. & Coffey, D.S. (1982). Biological properties of the nuclear matrix: steroid hormone binding. Recent Progress in Hormone Research, 38, 133-89.

Beach, D., Piper, M. & Shall, S. (1980a). Isolation of newly-initiated DNA from the early S-phase of the synchronous eukaryote, Physarum polycephalum. Experimental Cell Research, 129, 211-21.

Beach, D., Piper, M. & Shall, S. (1980b). Isolation of chromosomal origins of replication in yeast. Nature, 284, 185-7.

Beggs, J.D. (1978). Transformation of yeast by a replicating hybrid plasmid. Nature, 245, 104-9.

Bello, L.J. (1983). Differential transcription of early and late replicating DNA in human cells. Experimental Cell Research, 146, 79-86.

Blumenthall, A.B., Kriegstein, H.J. & Hogness, D.S. (1974). The units of DNA replication in Drosphila melanogaster chromosomes. Cold Spring Harbor Symposium in Quantitative Biology, 38, 205-23.

Bodnar, J.W., Jones, C.J., Coombs, D.H., Pearson, G.D. & Ward, D.C. (1983). Proteins tightly bound to HeLa cell DNA at the nuclear matrix attachment sites. Molecular and Cellular Biology, 3, 1567-79.

Broach, J.R., Li, Y.Y., Feldman, J., Jayaram, M., Abraham, J., Nasmyth, K.A. & Hicks, J.B. (1982). Localization and sequence analysis of yeast origins of DNA replication. Cold Spring Harbor Symposium in Quantitative Biology, 67, 1165-73.

Bryant, J.A. (1982). DNA replication and the cell cycle. In Encyclopaedia of Plant Physiology Volume 14B, eds. B. Parthier & D. Boulter, pp. 75-110. Berlin: Springer-Verlag.

Buongiorno-Nardelli, M., Micheli, G., Carri, M.T. & Marilley, M. (1982). A relationship between replicon size and supercoiled loop domains in the eukaryotic genome. Nature, 298, 100-2.

Butler, P.J.G. (1984). The folding of chromatin. CRC Critical Reviews in Biochemistry, 15, 57-91.

Chan, C.S.M. & Tye, B.-K. (1980). Autonomously replicating sequences in Saccharomyces cerevisiae. Proceedings of the National Academy of Sciences, USA, 77, 6329-33.

Chan, C.S.M. & Tye, B.-K. (1983). A family of Saccharomyces cerevisiae repetitive autonomously replicating sequences that have very similar genomic environments. Journal of Molecular Biology, 168, 505-23.

Clayton, D. (1982). Replication of animal mitochondrial DNA. Cell, 28, 693-705.

Dijkwel, P.A., Mullenders, L.H.F. & Wanka, F. (1979). Analysis of the attachment of replicating DNA to a nuclear matrix in mammalian interphase nuclei. Nucleic Acids Research, 6, 219-30.

Edenberg, H.J. & Huberman, J.A. (1975). Eukaryotic chromosome replication. Annual Review of Genetics, 9, 245-84.

Evans, T.E. (1982). Organization and replication of DNA in Physarum polycephalum. In Cell Biology of Physarum and Didymium, eds. H.C. Aldrich & J.W. Daniel, pp. 371-91. New York: Academic Press.

Franke, W.W., Scheer, U., Krohne, G. & Jarasch, E.-D. (1981). The nuclear envelope and the architecture of the nuclear peripherie. Journal of Cell Biology, 91, 39s-50s.

Funderud, S., Andreassen, R. & Haugli, F. (1978a). DNA replication in Physarum polycephalum: UV photolysis of maturing 5-bromodeoxyuridine substituted DNA. Nucleic Acids Research, 5, 3303-13.

Funderud, S., Andreassen, R. & Haugli, F. (1978b). Size distribution and maturation of newly replicated DNA through the S and G2 phases of Physarum polycephalum. Cell, 15, 1519-26.

Funderud, S., Andreassen, R. & Haugli, F. (1979). DNA replication in Physarum polycephalum: electron microscopic and autoradiographic analysis of replicating DNA from defined stages of the S period. Nucleic Acids Research, 6, 1417-31.

Goldberg, G.I., Collier, I. & Cassel, A. (1983). Specific DNA sequences associated with the nuclear matrix in synchronised mouse 3T3 cells. Proceedings of the National Academy of Sciences, USA., 80, 6887-91.

Gorman, J.A., Dove, W.F. & Warren, N. (1981). Isolation of Physarum DNA segments that support autonomous replication in yeast. Molecular and General Genetics, 183, 306-13.

Hanks, S.K., Gollin, S.M., Rao, P.N., Wray, W. & Hittleman, W.N. (1983). Cell cycle specific changes in the ultrastructural organization of prematurely condensed chromosomes. Chromosoma, 88, 333-42.

Hardman, N. & Gillespie, D.A.F. (1980). DNA replication in Physarum polycephalum. European Journal of Biochemistry, 106, 161-7.

Haugli, F., Andreassen, R. & Funderud, S. (1982). DNA replication in Physarum polycephalum: electron microscopic analysis of patterns of DNA replication in the presence of cycloheximide. Journal of Cell Biology, 95, 323-31.

Hay, R.T. & De Pamphilis, M. (1982). Initiation of SV40 DNA replication in vivo; location and structure of 5' ends of DNA synthesised in the ori region. Cell, 28, 767-79.

Hinnen, A., Hicks, J.B. & Fink, G.R. (1978). Transformation of yeast. Proceedings of the National Academy of Sciences, USA., 75, 1929-33.

Holmquist, G. & Goldman, M. (1983). Gene replication timing and chromosome structure. Journal of Cell Biology, 97, 112a.

Holt, C.E. (1980). The nuclear replication cycle in Physarum polycephalum. In Growth and Differentiation in Physarum polycephalum, eds. W.F. Dove & H.P. Rusch, pp.9-63. Princeton N.J.: Princeton University Press.

Hyde, J.E. (1982). Expansion of chicken erythrocyte nuclei upon limited micrococcal nuclease digestion. Experimental Cell Research, 140, 63-70.

Jalouzot, R., Briane, D., Ohlenbusch, H.H., Wilhelm, M.L. & Wilhelm, F.X. (1980). Kinetics of nuclease digestion of Physarum polycephalum nuclei in different stages of the cell cycle. European Journal of Biochemistry, 104, 432-31.

Kingsman, A.J., Clarke, L., Mortimer, R.K. & Carbon, J. (1979). Replication in Saccharomyces cerevisiae of plasmid pBR313 carrying DNA from the yeast trpI region. Gene, 7, 141-52.

Kornberg, A. (1980). DNA replication. San Francisco: Freeman.

Lau, Y. -F. (1983). Studies on mammalian chromosome replication. Experimental Cell Research, 146, 445-50.

McCready, S.J., Godwin, J., Mason, D.W., Brazell, I.A. & Cook, P.R. (1980). DNA is replicated at the nuclear cage. Journal of Cell Science, 46, 365-86.

Matthews, H.R. & Bradbury, E.M. (1982). Chromosome organization and chromosomal proteins in Physarum polycephalum. In Cell Biology of Physarum and Didymium, eds. H.C. Aldrich & J.W. Daniel, pp. 317-69. New York: Academic Press.

Matthews, H.R. & Waterborg, J.H. (1984). Reversible modifications of nuclear proteins and their significance. In The enzymology of post-translational modification of proteins, eds. R. Freedman & H.C. Hawkins, in press. London: Academic Press.

Mohberg, J. (1977). Nuclear DNA content and chromosome numbers throughout the life cycle of the Colonia strain of the myxomycete Physarum polycephalum. Journal of Cell Science, 24, 95-108.

Mohberg, J. (1982). Ploidy throughout the life cycle in Physarum polycephalum. In Cell Biology of Physarum and Didymium, eds. H.C. Aldrich & J.W. Daniel, pp. 253-72. New York: Academic Press.

Muldoon, J.J., Evans, T.E., Nygaard, O.F. & Evans, H.H. (1971). Control of DNA replication by protein synthesis at defined times during the S period in Physarum polycephalum. Biochimica et Biophysica Acta, 247, 310-21.

Mullenders, L.H.F., Eygenstein, J., Broen, A. & Wanka, F. (1982). Composition and DNA-binding properties of the nuclear matrix proteins from mammalian cell nuclei. Biochimica et Biophysica Acta, 698, 70-7.

Nagl, W. (1982). Nuclear chromatin. In Encyclopaedia of Plant Physiology Volume 14B, eds. B. Parthier & D. Boulter, pp. 1-45. Berlin: Springer-Verlag.

Smith, J.C. & Berezney, R. (1983). Dynamic domains of DNA polymerase alpha in regenerating rat liver. Biochemistry, 22, 3042-6.

Soeda, E., Arrand, J.R., Smolar, N. & Griffin, B.E. (1979). Sequence from early region of polyoma virus DNA containing viral replication origin and encoding small, middle and (part of) large T antigens. Cell, 17, 357-70.

Struhl, K., Stinchcomb, D.T., Scherer, S. & Davis, R.W. (1979). High-frequency transformation of yeast: autonomous replication of hybrid DNA molecules. Proceedings of the National Academy of Sciences, USA, 76, 1035-9.

Valenzuela, M.S., Mueller, G.C. & Dasgupta, S. (1983). Nuclear matrix-DNA complex resulting from EcoRI digestion of HeLa nucleoids is enriched for DNA replicating forks. Nucleic Acids Research, 11, 2155-64.

Van't Hof, J., Kuniyuki, A. & Bjerknes, C.A. (1978). The size and number of replicon families of chromosomal DNA of Arabidopsis thaliana. Chromosoma, 68, 269-85.

Vogelstein, B., Pardoll, D.M. & Coffey, D.S. (1980). Supercoiled loops and eukaryotic DNA replication. Cell, 22, 79-85.

Vogt, V.M. & Braun, R. (1977). The replication of ribosomal DNA in Physarum polycephalum. European Journal of Biochemistry, 80, 557-66.

Wanka, F. (1984). Cell cycle studies of DNA association with the nuclear matrix. In The Nuclear Matrix, ed. R. Berezney, in press. New York: Plenum.

Wille, J.J. & Kauffman, S.A. (1975). Premature replication of late S period DNA regions in early S nuclei transferred to late S cytoplasm by fusion in Physarum polycephalum. Biochimica et Biophysica Acta, 407, 158-73.

Wille, J.J. (1977). Preferential binding of S phase proteins to temporally characteristic units of replication in Physarum polycephalum. Nucleic Acids Research, 4, 3143-54.

ENZYMIC CONTROLS OF DNA REPLICATION

V.L. Dunham and J.A. Bryant

INTRODUCTION

As indicated in other chapters in this volume, the controls involved in the regulation of the plant cell cycle are varied, may function at several phases in the cycle, and may even be demonstrated at unique times within a given phase of the cycle. The various controls of the plant cell cycle cover a gamut of possibilities ranging from "general" controls, such as cell size, nutrition and the effects of plant growth regulators on protein, RNA and DNA synthesis, to controls over the synthesis and/or activity of specific proteins. For example, numerous studies have indicated that an inhibition of protein synthesis may inhibit DNA synthesis and consequently, movements of cells through the cycle (Nurse & Thuriaux, 1977; Nasmyth, Nurse & Fraser, 1979). In addition, specific stages within the S-phase may be delayed in the presence of an inhibitor of protein synthesis. For example, cycloheximide has been shown to inhibit nascent DNA maturation in a concentration-dependent manner (Schvartzman & Van't Hof, 1982). Since DNA replication during the S-phase of the cell cycle requires numerous proteins, the synthesis and/or activity of some of these proteins may be points of control of the plant cell cycle or at least function in regulating DNA replication. Although there have been reports of numerous soluble factors which appear to regulate the S-phase of the cell cycle (reviewed by Bryant, 1976) little progress has been made in characterizing them. This paper will, therefore, focus on those proteins considered to function at, or near the replication fork and to be directly involved in DNA replication (Fig. 1). The discussion will include: (A) characteristics of the proteins, (B) possible functions and mechanisms for their recognition of DNA, (C) intracellular concentrations and activity of enzymes during the cell cycle and during development and (D) possible regulation by plant growth regulators.

Fig. 1. Schematic representations of replication forks in
(a) prokaryotes
(b) eukaryotes

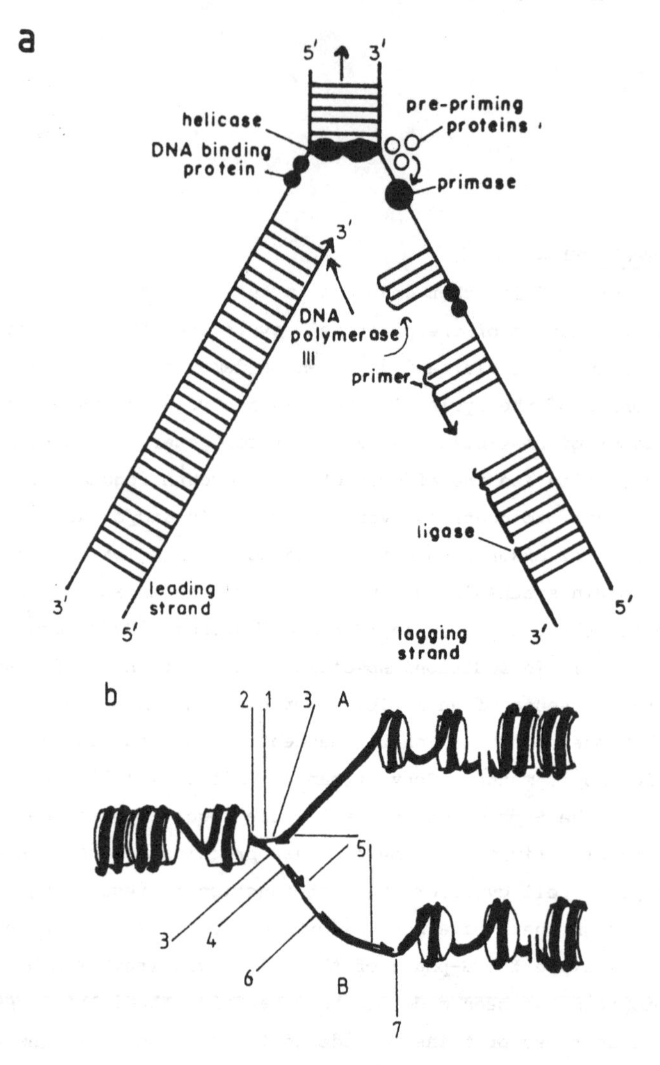

1 Helicase
2 Topoisomerase
3 DNA-binding protein
4 Primase
5 DNA polymerase
6 Ribonuclease H
7 DNA ligase

A. Leading strand
B. Lagging strand

PROTEINS OF DNA REPLICATION

As illustrated in Fig. 1a, replication of prokaryotic DNA requires the presence and activity of numerous proteins. Although not all of the proteins illustrated in Fig. 1 have been found in plants, or for that matter, in any eukaryote (Fig. 1b), we will use the illustration (Fig. 1a & b) for a guide to a description of the proteins involved in the replication of plant DNA and to serve as a comparison as to proposed functions of the proteins in plants. The proteins are dealt with for the most part in the order in which they are presumed to function.

Single-stranded DNA binding protein (SSB)

The A.T.-rich nature of the replication origins so far described (see Waterborg & Shall, this volume) indicates that these sequences are prone to "breathing", i.e. transient single-strandedness. An early event in replication is thus likely to be the stabilization of this single-stranded state. Further, such stabilization must also be a feature of the moving replication fork. SSBs have been isolated and characterized from several eukaryotic cells. The SSB from calf thymus displays the typical characteristics of these proteins including preferential binding to single-stranded DNA, reduction of T_m, (i.e. the temperature at which DNA is half-denatured) and the stimulation of DNA polymerase-α (Herrick & Alberts, 1976a, b). These properties support the idea that the SSBs function in DNA replication by stabilizing single-stranded regions at the replication fork.

SSBs have been isolated from the fungus *Ustilago* (Banks & Spanos, 1975) and from one higher plant, *Lilium* (Hotta & Stern, 1971, 1979). The protein from *Ustilago* has a molecular weight of 20,000 and is capable of lowering the T_m of poly-d(A-T) - d(A-T) by 50°C. In *Lilium*, SSB was found only in meiotic cells (developing pollen grains) and reached maximum cellular concentrations at the time of genetic recombination (crossing over). These findings suggest that SSBs may also be involved in recombination events as well as in replication. In addition, recent evidence from Novikoff hepatoma cells indicates the presence of a SSB that specifically stimulates DNA polymerase-β and allows an increased rate and extent of DNA synthesis (Koerner & Meyer, 1983). Since polymerase-β is believed to be involved in some types of DNA repair (Bryant, 1982; Smith & Okumoto, 1984) these data indicate that this SSB may alter the conformation of the template strand to facilitate DNA repair synthesis, as well as replication.

DNA Primase

DNA replication in vivo requires the existence of an oligoribonucleotide primer. Recent investigations of the subunit structure of DNA polymerase-α in eukaryotes have led to the discovery of an associated DNA primase activity in Drosphila embryos (Kaguni et al., 1983), and in mammals (Yagura et al., 1983). The DNA polymerase-primase complex of Drosophila has been characterized and consists of at least three sub-units α, β, and γ with molecular weights of 182,000, 60,000 and 50,000 respectively. The α subunit (DNA polymerase-α activity) is not required for DNA primase activity. In vertebrates the situation is less clear. Primase is certainly associated with the catalytic subunit, but whether or not it is part of the same subunit remains to be determined (evidence is reviewed in Kaguni et al., 1983, and Yagura et al., 1983). Recent investigations of purified DNA polymerase-α in higher plants have indicated that no DNA primase activity is found in polymerase preparations (Graveline, Tarrago-Litvak, Castroviejo & Litvak, 1984; J.A. Bryant & P.N. Fitchett, unpublished data). In fact, a DNA primase activity has been observed in wheat (Triticum aestivum) embryos, separate from DNA polymerase-α, and found to have a molecular weight of 70,000 (Graveline et al., 1984).

DNA Polymerase-α

Numerous studies relating DNA polymerase activity to increased cell proliferation indicate that of the multiple DNA polymerase present in higher plants, DNA polymerase-α is probably the replicative enzyme (Bryant, 1980, 1982). In general, DNA polymerase-α is a protein of high molecular weight. It is prone to two forms of breakdown: (i) loss of non-catalytic subunits and (ii) proteolyis of the major catalytic subunit (Table 1). For example, the enzyme from spinach (Spinacea oleracea) leaves has been reported to have a molecular weight of 160,000 (two subunits of ca. 80,000), but a catalytically active fragment of 12,000 is generated by proteolysis during extraction (Misumi & Weissbach, 1982). Further, in view of the sizes currently assigned to the catalytic subunits of DNA polymerase-α in other eukaryotes (140 to 182,000), even the 80,000 subunits may represent partial proteolysis products. In addition, polymerase-α typically aggregates to high molecular weight forms in solutions of low ionic strength. In pea (Pisum sativum) the catalytic subunit of DNA polymerase-α has a molecular weight of 180,000 (or higher) and breaks down through a series of sizes including 140,000, 110,000, 75,000 and 45,000 (Chivers & Bryant, 1983; J.A. Bryant &

Table 1: DNA Polymerase-α Enzymes in Higher Plants

	*Mol.wt (kdal) and/or	sed. coeff.	KCL,mM	Optima Mg, mM	pH	Inhibition by NEM	aphidicolin
1. *Beta vulgaris* (beet)	113	7S		15	7.5	+	
2. *Pisum sativum* (pea)	180		0-25	15	8.1	+	+
3. *Glycine max* (soybean)				6	7.5		+
4. *Vinca rosea* (periwinkle)	105+		50	6-15	7.5	+	
5. *Oryza sativa* (rice)		7S	0	5-10	7.2-8.4	+	+
6. *Brassica rapa* (turnip)			0-25	15	8.0	+	+
7. *Triticum sp.* (wheat)	110	(pol 'B')	0	5	7.0	+	
8. *Spinacia oleracea* (spinach)	160++		0	10	7.0-8.5	+	+
9. *Nicotiana tabacum* (tobacco)		7-10S	50 (NaCl)	7	8.3	+	

* These molecular weights are, in most instances, the weights which may be assigned to the catalytic subunit. The latter may well be associated with other polypeptides to form a large, complex enzyme.

\+ This is made up of two subunits, one of ca. 70,000 and one of ca. 35,000.

++ This is made up of two subunits, each of ca. 70-80,000.

References
1. Tymonko & Dunham (1977); 2. Stevens & Bryant (1978), Chivers & Bryant (1983), J.A. Bryant & P.N. Fitchett, (unpublished data); 3. D'Alesandro, Jaskot & Dunham (1980); 4. Gardner & Kado (1976); 5. Amileni, Sala, Cella & Spadari (1979), Sala, Amileni, Parisi & Spadari (1980); 6. Dunham & Bryant (1981), Thomas, Bryant, Dunham & Hull (1983); 7. Castroviejo, Tarrago-Litvak & Litvak (1975), Castroviejo, Tharaud, Tarrago-Litvak & Litvak (1979); 8. Misumi & Weissbach (1982); 9. Srivastava (1974).

P.N. Fitchett, unpublished data). The enzyme also possesses at least two additional non-catalytic subunits (J.A. Bryant & P.N. Fitchett, unpublished data).

Although polymerase-α is widely believed to be the replicative enzyme, DNA polymerase-γ (see Bryant, 1982) has been reported to be the replicative enzyme in wheat (*Triticum aestivum*) (Graveline *et al.*, 1984). This enzyme has three subunits and is the only wheat polymerase that has the ability to recognize a primer. These findings are in contrast to the information from most higher plants and from a wide range of other eukaryotic organisms. Further work is necessary to resolve the situation. It is notable, for example, that wheat embryos also have two DNA polymerase-α-like activities (enzymes 'B' and 'C2') which have an approximate molecular weight of 110,000 (Castroviejo, Tharaud, Tarrago-Litvak & Litvak, 1979). If polymerase-γ is the replicative enzyme in wheat nuclei, what is the role of polymerase-α in this species?

DNA polymerase-α has been identified in soluble preparations by its sensitivity to low concentrations of N-ethylmaleimide (NEM) and more specifically, to aphidicolin (a known specific inhibitor of DNA polymerase-α in animal systems). In general, most DNA polymerase-α activity is inhibited at NEM concentrations of 1 mM or less and by aphidicolin at 30 μM or less (Table 1). The enzyme has a Mg^{2+} optimum at 10-15 mM, is generally inhibited by KCl and prefers an activated (i.e. gapped with deoxyribonuclease) template-primer system. Most of the DNA polymerase-α activities isolated and purified from higher plants do not have associated nuclease activity. The enzyme is also characterized by its ease of solubility from DNA and leakage from isolated nuclei. In most higher plants, 70-90% of the DNA polymerase-α activity is found in the soluble fraction of cell homogenates. The small amount of DNA polymerase-α remaining in the nucleus may be extracted with 0.35 M KH_2PO_4 (Misumi & Weissbach, 1982) and may be associated with the nuclear matrix.

Ribonuclease H

The existence of oligonucleotide primers at the 5' ends of Okazaki fragments and the fact that most eukaryotic DNA polymerases lack nuclease activity, lead to the necessity for an enzyme to remove these primers. Ribonuclease H is an enzyme that specifically degrades RNA in a DNA/RNA hybrid molecule. The enzyme has been detected in a number of eukaryotic organisms including *Saccharomyces* (Wyers, Sentenac & Fromageot,

1973), Ustilago (Banks, 1974) and carrot (Daucus carota)(Sawai, Sugano & Tsukada, 1978). In general, the enzyme has a molecular weight of 70-150,000, requires divalent cations for activity and is an endonuclease. Although the activity of the enzyme increases in carrot cells actively engaged in DNA synthesis, there is to date, no direct evidence for its involvement at the replication fork.

DNA Ligase

DNA ligase is an enzyme that functions in joining fragments of newly synthesized DNA by forming a phosphodiester bond between the 3' OH end of one fragment and the 5' P end of the other fragment (nick sealing, Fig. 2). In prokaryotic cells, the enzyme is NAD-dependent, whereas in eukaryotes it is ATP-dependent (Kornberg, 1980). Although DNA ligase activity has been identified in several plants, the enzyme has not been purified and characterized. The enzyme has been shown to be present in several plants including pea, soybean, spinach and cucumber (Kessler, 1971). The activity appears to be localized in the nucleus and is dependent on the presence of ATP and Mg^{2+}. In the microspores of Lilium, on the other hand, the enzyme has been reported as soluble (Howell & Stern, 1971). The characteristics and function of the enzyme in replication, repair and recombination in higher plants remains to be elucidated. Indeed, despite having been detected in plants as long ago as 1971, almost no progress has been made on plant DNA ligase since then. In pea (Pisum sativum) the enzyme appears to be unstable (P.P. Daniel & J.A. Bryant, unpublished data) which may be the reason for lack of data.

Helicase

Helicase is a DNA-dependent ATPase that uses ATP to destabilize the base pairs in advance of the replicating fork and thus serves to unwind the double helix (Kornberg, 1982). Although numerous helicases have been characterized in prokaryotes, including at least three in E. coli, only one has been described in plants. In Lilium, a helicase has been identified which has molecular weight of 130,000, but its exact function has not been reported (Hotta & Stern, 1978).

Gyrase and Topoisomerase

The progression of the replication fork by the separation of two intertwined strands of the double helix automatically leads to the occurrence of positive supercoiling (i.e. overwinding of the helix). In prokaryotes this

Fig.2. Mode of action of DNA ligase

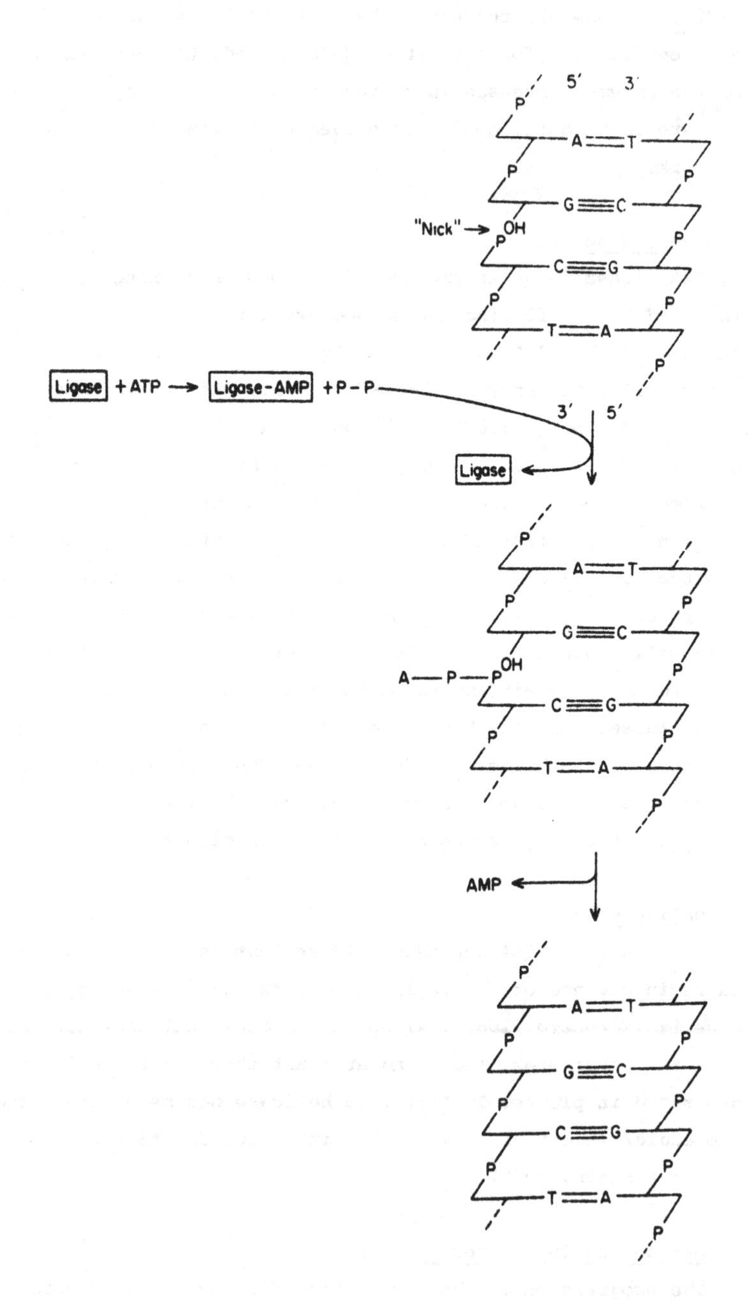

problem appears to be overcome by the action of an enzyme called gyrase, which inserts negative supercoils (i.e. causes helix underwinding) in front of the replication fork. This activity is ATP-dependent and in addition to counteracting the positive supercoiling caused by fork movement, also allows the binding of several of the proteins which act at the replication fork (Kornberg, 1980).

Gyrase is in fact a special type of topoisomerase, a class of enzymes which change the degree of supercoiling of DNA. Although there is no clear evidence for gyrase *per se* in eukaryotes, two other topoisomerases have been isolated from eukaryotic cells. Type I topoisomerases relax supercoiling by a mechanism which involves nicking one strand, rotation and re-sealing. Type II enzymes can bring about an ATP-dependent positive supercoiling of relaxed DNA via a mechanism which involves nicking and re-sealing of both strands (Kornberg, 1980). In terms of aiding the progression of the replication fork, current evidence, albeit sparse, favours the participation of topoisomerase I. The type I topoisomerases isolated from yeast (*Saccharomyces cerevisiae*) (Durnford & Champoux, 1978) and from wheat (*T. aestivum*) embryos (Dyan, Jendrisak, Hager & Burgess, 1981) display typical eukaryotic characteristics (see Kornberg, 1982) including lack of a requirement for magnesium and for ATP, the ability to relax both positively and negatively supercoiled DNA, and a molecular weight of ca. 110,000. Type 1 topoisomerase activity has also recently been identified in pea (*P. sativum*) (J.A. Bryant & P.N. Fitchett, unpublished data). The enzyme activity is again independent of Mg^{2+} and ATP; it requires NaCl at an optimum concentration of 100 mM and is stimulated by phosphate ions (Fig. 3).

Endodeoxyribonuclease

Although current opinion favours topoisomerase as being involved in relaxation of supercoils, earlier evidence had implicated single-strand-specific endonuclease in this activity (see discussion in Bryant, 1980, 1982). Because of their single-strand specificity, such enzymes nick double-stranded DNA in A.T.-rich regions, particularly when the tendency of these regions to exhibit transient denaturation (i.e. "breathe") is accentuated by supercoiling. Such an enzyme could therefore nick one DNA strand at intervals, allowing free rotation of a stretch of DNA. Re-ligation would follow after passage of the replication fork. Single-strand-specific endodeoxyribonucleases have been detected in a number of plants (Bryant, 1980, 1982). Typical is the enzyme from pea (*P. sativum*) which is chromatin-bound, exists as a single polypeptide chain of molecular weight 18,000 and shows some correlation with DNA replication (Jenns & Bryant, 1978; Weir, 1983; A.F. Weir & J.A. Bryant, unpublished data).

The existence of both topoisomerase I and single-strand-specific endodeoxyribonuclease within the nuclei of the same plant (pea) suggests that these two enzymes have different roles. If indeed topoisomerase is involved in relaxation of supercoiling, then a possible role for the endonuclease is in post-replication repair (Weir, 1983).

INTERACTION OF PROTEINS WITH DNA DURING REPLICATION

Recognition of specific regions of the double helix by proteins

Fig.3. Action of pea chromatin topoisomerase I on supercoiled plasmid DNA. The numbers at the top of the gel lanes represent incubation time in minutes. As incubation times are increased, the supercoiled DNA is progressively converted to the relaxed form via a specific series of topoisomers.

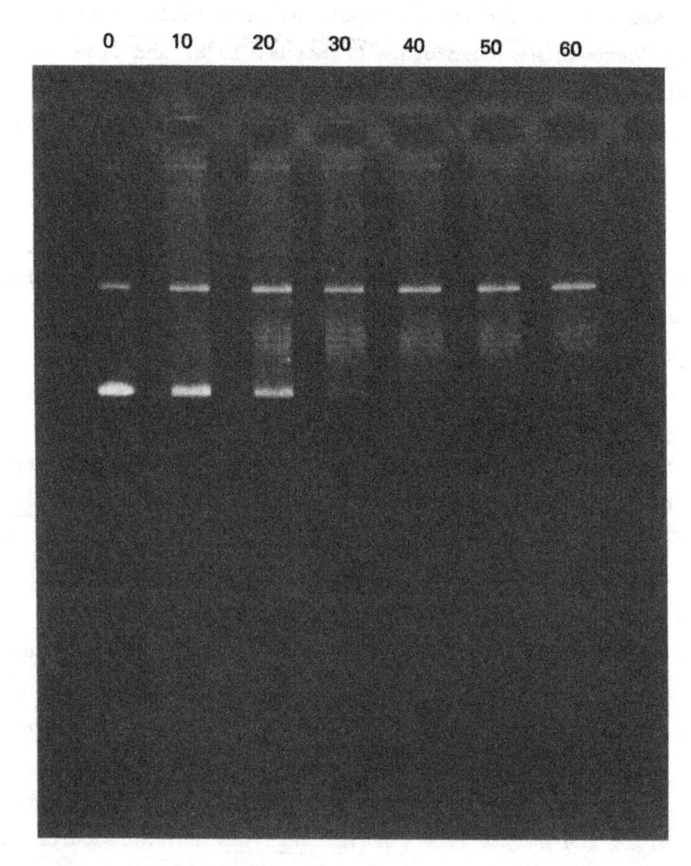

involved in DNA replication is essential for both the initiation of replication and for maintaining the proper rate of fork movement. The following discussion will focus on information concerning the recognition of origins of replication and possible interaction and organization of the proteins involved in DNA replication.

Replication origins

It is apparent that the sequence of bases at the origin of DNA replication in prokaryotes is essential for the proper recognition by replicative proteins (Kornberg, 1982). For example, a consensus sequence of CACTAT is present in the origins of a number of viral DNAs and has been reported to act as a recognition site for proteins involved in DNA replication as well as transcription.

The origins of DNA replication in eukaryotes differ from those of prokaryotes in that they are multiple (many thousands of copies in a typical plant genome) and are not re-initiated before complete replication of the chromosome has occurred. In different plant species these origins of DNA replication are between 20 and 150 kb apart (see Van't Hof and Francis, Kidd & Bennett, this volume). In addition, in pea there appears to be a correlation between the number of replicons used and the rate of DNA synthesis (Van't Hof, 1976a, b). Autonomously replicating sequences (ars) have been shown to exist in eukaryotic genomes, including yeast, plant and mammalian DNAs, as detected by their ability to permit recombinant plasmids to replicate in yeast cells (see Waterborg and Shall, this volume). In yeast the ars appear in about 400 places in the genome, a number similar to the number of origins observed (Kornberg, 1982). In mammals, yeast and fission yeast, the ars are A.T.-rich, with strong sequence homology between organisms (e.g. Montiel *et al.*, 1984). Sequences of the origins for plant DNA replication are presently being investigated, but recognition mechanisms by proteins involved in replication remain to be studied. The expected complexity of these interactions is reflected by Arthur Kornberg's statement, "The promotional signals for eukaryotic replication may eventually resemble in variety the increasing complex assortment of signals and factors regulating the starts of eukaryotic transcription".

Protein-protein interactions

Not only must proteins involved in DNA replication recognize specific regions of DNA such as origins, but an increasing number of reports indicate that these proteins often recognize DNA regions by their specific interaction with other proteins to form complexes, perhaps initially overlooked due to dissociation during purification procedures. The prokaryotic primosome, responsible for the laying down of RNA primers for DNA replication on the lagging strand especially (Fig. 1a), consists of numerous proteins including pre-priming proteins and primase. The replisome is a more inclusive complex that may be composed of a dimeric DNA polymerase, primosome and one or more helicases (Kornberg, 1982).

Evidence for the involvement of protein complexes in the replication of DNA in eukaryotes has focused on the existence of "replitase", a large complex of at least six enzymes involved in DNA metabolism, apparently found in the nuclear region only during the S-phase of hamster fibroblast cells (Reddy & Pardee, 1980). This complex of enzymes is not present in quiescent (GO) and G1-phase cells. The complex has been shown to include DNA polymerase-α and several enzymes involved with dNTP synthesis such as ribonucleotide reductase and thymidylate synthase (Reddy, 1982). The proposed function of the complex during S-phase includes DNA replication and the metabolic channeling of nucleotides for on-site synthesis of dNTPs. There is evidence consistent with this idea from yeast cells, where the replicative DNA polymerase only associates with the nucleus during S-phase (Tsuchiya, Kimara, Miyakawa & Fukui, 1984).

Evidence from higher plants as to the existence of these complexes is presently lacking. It is interesting to note that topoisomerase activity appears to be chromatin-bound in pea (P. sativum) and therefore could be part of a nuclear complex (P.N. Fitchett & J.A.Bryant, unpublished data). The finding that DNA primase activity is separate from DNA polymerase-α in wheat (T. aestivum) embryos (Graveline et al., 1984) does not preclude the possibility of direct interaction in a complex during the S-phase of the cell cycle. Further investigation of the intra-cellular location of several of these proteins, especially during the G1 and S-phase of the plant cell cycle, is required.

Nuclear matrix involvement in DNA replication

Labelling of DNA during the cell cycle of Physarum has indicated that origins of replicons or DNA sites closely associated with them are attached to the nuclear matrix during the entire nuclear cycle and that initiation of DNA replication occurs at these same sites in the next S-phase

(Aelen, Opstelten & Wanka, 1983). However, autonomous replicating sequences from human cells are reported to map at sites in the DNA, apparently remote from attachment sites in the HeLa nuclear matrix (Cook & Lang, 1984), although Goldberg, Collier & Cassel (1983) detected DNA with properties similar to autonomous replicating sequences attached to the nuclear matrix in mouse cells synchronized at the G1/S boundary (see also Waterborg & Shall, this volume). Interaction of replicating proteins and DNA with the nuclear membrane and matrix of higher plant cells remains to be investigated.

Modification of proteins involved in DNA replication

Post-translational modification of proteins involved in DNA replication may play a role in regulating DNA synthesis. The SSB isolated from *Lilium*, for example, has been shown to be a substrate for cyclic-AMP-dependent protein kinase (Hotta & Stern, 1979). Proteins isolated from soybean (*Glycine max*) nuclei and incubated with ATP, to allow for phosphorylation by endogenous, chromatin-bound protein kinases, have been shown to alter DNA polymerase-α-like activity (Dunham & Yunghans, 1977). Of five protein fractions investigated, one fraction inhibited DNA polymerase activity almost two-fold more following incubation with ATP, whereas two protein fractions which stimulated DNA synthesis had a greater stimulatory effect following incubation with ATP.

REGULATION OF ENZYMES IN RELATION TO CELL PROLIFERATION AND PLANT DEVELOPMENT

Although situations may occur in which plant DNA synthesis is not closely correlated with cell division (repair, unscheduled DNA synthesis, endo-cycles), the replication of DNA is usually confined to the S-phase of the cell division cycle and hence, in higher plants to meristematic cells. The following discussion will attempt to correlate the synthesis and activity of specific enzymes with cell proliferation, and with certain aspects of plant development.

Increased enzyme activity correlated with cell proliferation

DNA polymerase activity is high in cells that are rapidly dividing, but few studies have delineated the types of DNA polymerase involved. In cultures of rice (*Oryza sativa*), sycamore (*Acer pseudoplatanus*) and Virginia creeper (*Parthenocissus tricuspidata*), the major DNA polymerase activity (identified as DNA polymerase-α) is low in stationary phase, increases several fold with the onset of cell proliferation, and decreases at the end of the logarithmic phase (Amileni, Sala, Cella & Spadari, 1979). Mixing of extracts of late stationary phase and extracts from cells in the logarithmic phase with the enzyme did not alter enzyme activity, indicating that the enzyme was degraded or inactivated at the end of log phase and not inhibited by a molecule synthesized in older cells. Further evidence for the involvement of DNA polymerase-α in DNA replication in cultured plant cells comes from experiments with aphidicolin, a specific inhibitor of DNA polymerase-α (Sala *et al.* 1981). Addition of aphidicolin to cultured carrot (*D. carota*) cells over a 24 hour period caused the cycling cells to build up at the G1/S boundary of the cell cycle. Removal of aphidicolin resulted in a resumption of nuclear DNA synthesis and the synchronous movement of the cells throughout the S-phase (Sala *et al.*, 1983). These data further support the idea that DNA polymerase-α is the enzyme involved in nuclear DNA replication and that it is required during the S-phase of the cell cycle.

Increased levels of DNA polymerase-α activity have also been shown to be associated with meristematic cells. For example, Bryant, Jenns & Francis (1981), using serial sections of roots from 5-day-old pea (*P. sativum*) seedlings, have shown that regions of the pea root that exhibit the highest rate of DNA replication (based on mitotic indices and incorporation of H-thymidine into DNA) also exhibit the highest levels of DNA polymerase-α (Fig. 4). Enzyme activity is reduced in regions of cell differentiation and maturation. Mixing extracts from different regions of the root with the enzyme indicates the absence of any regulatory molecules specifically produced at some stage of development. DNA polymerase-β, the chromatin-bound enzyme, was by contrast, present at its highest levels in regions of elongation and maturation. Similar increased levels of DNA polymerase-α activity in meristematic cells have been observed in the "hook" region of etiolated 4-day old soybean (*G. max*) hypocotyls (V.L. Dunham, unpublished data). Although these data show clear evidence for fluctuation in DNA polymerase-α in relation to cell age, there is currently no evidence that the activity fluctuates during the cell cycle within meristematic cells: i.e. total activity is as high during G1 as in S (C.M. Thomas, J.A. Bryant & D. Francis, unpublished data).

Few other enzymes implicated in DNA replication have been investigated in relation to cell proliferation. However, ribonuclease H increases in activity in the log phase of cultured carrot (*D. carota*) cells in parallel with a similar increase in DNA polymerase-α activity (Sawai *et* al., 1978).

Increased DNA polymerase activity associated with seed germination

Uptake of water into seeds triggers an array of events, one being an increased rate of DNA synthesis as measured by incorporation of ^{3}H-thymidine. In wheat (T. aestivum), this incorporation begins approximately 12 hours after the onset of imbibition (Castroveijo, Tharaud, Tarrago-Litvak & Litvak, 1979). DNA polymerase activity, however, could be isolated from ungerminated wheat seeds and increased during germination (Mory, Chen &

Fig.4. Relationship between DNA labelling index, mitotic index and DNA polymerase-α activity in roots of pea seedlings. Enzyme units are nmol dTMP incorporated per mg DNA per hour.

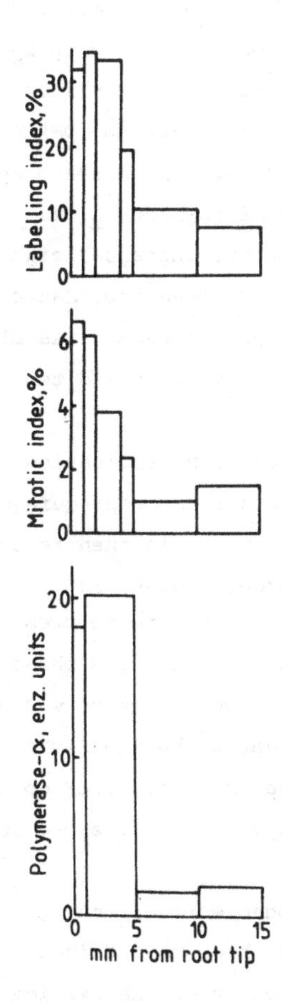

Sarid, 1975). In addition, the eight-fold increase in DNA polymerase activity following imbibition was apparently dependent on de novo biosynthesis of the enzyme (the increase was inhibited by blasticidin S). Several different DNA polymerase activities have been isolated and partially purified from soluble fractions of wheat seedlings (Castroviejo et al., 1979; Graveline et al., 1984). Two of the activities (polB and C2) were similar to polymerase-α, present in ungerminated seeds and increased in activity over the first 10 hours of germination.

The specific DNA polymerases involved in increased DNA biosynthesis in germinating tissues have also been identified by analysis of soluble (α-like) and chromatin-bound (β-like) enzymes from pea (P. sativum seedlings (Robinson & Bryant, 1975). Although net DNA synthesis in pea seeds was observed approximately 30 hours after the onset of germination, both soluble (α) and chromatin-bound (β) DNA polymerases were present as early as six hours. Polymerase-β reached maximum activity by 23 hours (less than a three-fold increase) whereas polymerase-α continued to increase through 47 hours (over four-fold increase). In addition, DNA polymerase-α activity was generally 50 times that observed for polymerase-β activity. DNA polymerase-α, present in the soluble fraction of soybean (G. max) embryos as early as six hours after the onset of imbibition, increased eight-fold (units/gm dry weight of embryo) over 48 hours (D'Alesandro, Jaskot & Dunham, 1980) (Fig. 5). The chromatin-bound enzyme, DNA polymerase-β, was also present at low levels at 6 hours after imbibition but only increased two-fold over the same two day period.

In the early stage of seed germination in broad bean (Vicia faba), 96% of the cells which entered the first post-quiescence cell cycle were in G1 phase. The first wave of DNA synthesis then reached a peak between 30-40 hours after uptake of water (Jakob & Bovey, 1969). A DNA polymerase-α like enzyme has been observed in this system to be present at all stages of development and to increase two-fold during S-phase (Hovemann & Follman, 1979). Interestingly, ribonucleotide reductase was detected only 30-32 hours after water uptake and was thought to be the limiting enzyme for DNA replication. The regulation of both the intra-cellular concentration and the activity of this enzyme is quite complex and may serve as a major control point in the cell cycle.

In addition to ribonucleotide reductase and DNA polymerase mentioned above, another enzyme which may be involved with DNA synthesis is known to increase prior to the onset of DNA replication during germination.

Elevated levels of endodeoxyribonuclease have been shown to be associated with the onset of DNA synthesis in pea (*P. sativum*) embryos (Jenns & Bryant, 1978). An unidentified inhibitor may be associated with the enzyme prior to the observed increase in activity.

Changes in DNA polymerase activity in the wound response

The excision and then aerobic incubation of slices or discs of quiescent plant storage tissue results in a vastly increased metabolic activity, including elevated rates of protein and RNA synthesis, and under appropriate conditions, re-initiation of DNA replication (Bryant, 1976). The onset of DNA replication is partially synchronous in explants of potato (*Solanum tuberosum*) tuber (Watanabe & Imaseki, 1973) and of artichoke (*Helianthus tuberosus*) tuber (Harland, Jackson & Yeoman, 1973). Increased DNA polymerase activity is observed in both systems, and in potato has been

Fig. 5. Changes in activity of DNA polymerase-α(O) and DNA polymerase-β(O) during germination of soybean. Enzyme units are pmol dTMP incorporated per 10 min.

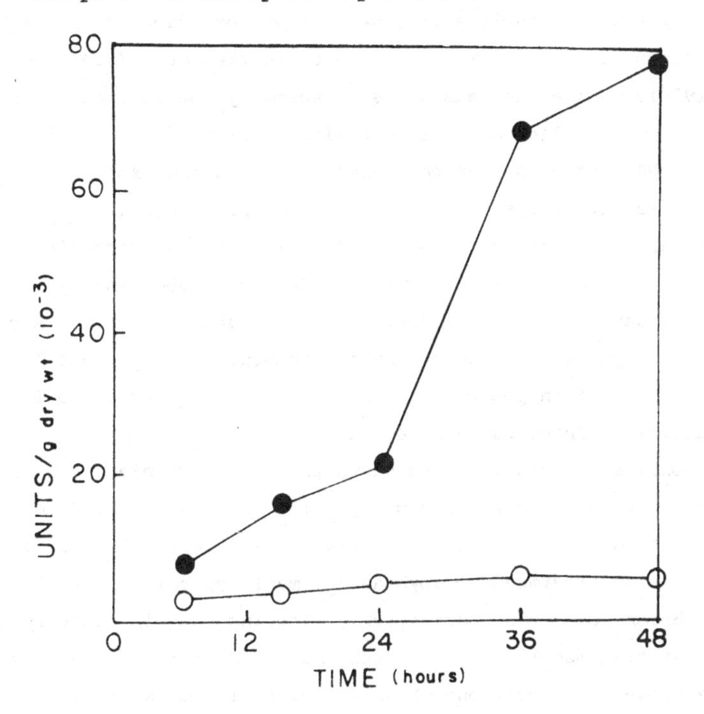

shown to be dependent on protein synthesis (Watanabe & Imaseki, 1977). The DNA polymerase in artichoke has not been extensively characterized, but in potato, the enzyme appears to be DNA polymerase-α, based on its observed sedimentation coefficient of 5.3S. DNA polymerase (chromatin-bound) in slices of sugar beet (Beta vulgaris) root reaches a maximum at 30 hours after excision and declines again by 48 hours (Dunham & Cherry, 1973). The enzyme exists as multiple forms which increase in number and activity during incubation. More recent work has indicated the presence of both polymerase-α and polymerase-β in these preparations (Tymonko & Dunham, 1977), but because of the tendency of the enzymes to form aggregates, it is difficult to interpret changes in the multiple activities as being due to one or other of these polymerase species.

Effects of plant growth regulators on enzymic control

Most plant growth regulators have been shown to alter DNA synthesis under a variety of conditions. This alteration may possibly be expressed via an indirect or direct effect on DNA polymerase activity or on other enzymes of DNA replication. Several workers have shown that DNA synthesis increases 12-18 hours following auxin treatment (reviewed by Trewavas, 1976). Similar increases, as detected by the incorporation of ^{3}H-thymidine, occur following treatment with cytokinins (Kende, 1971). Decreases in DNA synthesis have been observed following exposure of Lemna to abscisic acid (Stewart & Smith, 1972). Exposure of pea (P. sativum) seedlings to abscisic acid results in a decrease in DNA synthesis as measured by incorporation of ^{3}H-thymidine (Chivers, 1984) and by measurement of labelling index (D. Francis & J.A. Bryant, unpublished data). A similar decrease in DNA synthesis, detected by the incorporation of labelled precursors, was observed in potato (S. tuberosum) tuber discs following treatment with ethylene (Sato, Watanabe & Imaseki, 1976).

Only a few studies have shown the effect of plant growth regulators on the activity and synthesis of specific enzymes involved in DNA replication. Chromatin-bound DNA polymerase was observed to increase in activity following treatment of soybean (G. max) hypocotyls with 2,4D (Leffler, O'Brien, Glover & Cherry, 1971). The increase in activity (30%, 12 hours after treatment) coincided with induced mitosis in the tissue. Under these isolation conditions, however, most of the DNA polymerase-α activity would have been solubilized from the chromatin. In the artichoke tuber explant system described above, the onset of DNA replication and the associated increase in DNA polymerase are dependent on the presence of auxin (Harland *et al.*, 1973). However, as mentioned previously, it is not clear which DNA polymerase is involved.

In a soluble fraction obtained from the apices of etiolated pea seedlings, DNA polymerase activity was decreased following ethylene treatment (Apelbaum, Sfakiotakis & Dilley, 1974). Based on the properties of the enzyme and present knowledge of the types of polymerase in soluble extracts of pea, the decrease in activity was probably due to a decrease in the level of DNA polymerase-α activity. Although abscisic acid inhibits DNA synthesis in the same species, neither the synthesis nor activity of DNA polymerase-α and -β were affected by exposure to the growth regulator (Chivers, 1984).

CONCLUSIONS

Despite the characterization of a number of enzymes, particularly DNA polymerase, we are still a long way from a clear understanding of the nature and role of the proteins involved in replication of plant DNA. This in turn means that knowledge of the regulation of DNA replication at the biochemical level remains scant. Although there are some data on changes in activities of certain enzymes during plant development, or in response to plant growth regulators, the role of these changes in the tightly coordinated regulation of growth and development is not known.

ACKNOWLEDGEMENTS

We thank the following for financial support of our own research: U.S.A.: N.S.F., S.U.N.Y. Res. Foundation; U.K.: A.F.R.C., S.E.R.C., Royal Society, Nuffield Foundation, Leverhulme Trust, Unilever Research plc, Shell Research plc.

REFERENCES

Aelen, J.M.A., Opstelten, R.J.G. & Wanka, F. (1983). Organization of DNA replication in *Physarum polycephalum*. Attachment of origins of replicons and replication forks to the nuclear matrix. *Nucleic Acids Research* 11, 1181-95.

Amileni, A., Sala, F., Cella, R. & Spadari, S. (1979). The major DNA polymerase in cultured plant cells: partial purification and correlation with cell multiplication. *Planta* 146, 521-7.

Apelbaum, A., Sfakiotakis, E. & Dilley, D.R. (1974). Reduction in extractable DNA polymerase activity in *Pisum sativum* seedlings by ethylene. *Plant Physiology* 54, 125-8.

Banks, G.R. (1974). A ribonuclease H from Ustilago maydis. Properties, mode of action and substrate specificity of the enzyme. European Journal of Biochemistry 47, 499-507.

Banks, G.R. & Spanos, A. (1975). The isolation and properties of a DNA-unwinding protein from Ustilago maydis. Journal of Molecular Biology 93, 63-77.

Bryant, J.A. (1976). The cell cycle. In Molecular aspects of gene expression in plants, ed. J.A. Bryant, pp. 177-216. London & New York: Academic Press.

Bryant, J.A. (1980). Biochemical aspects of DNA replication, with particular reference to plants. Biological Reviews, 55, 237-84.

Bryant, J.A. (1982). DNA replication and the cell cycle. In Encyclopedia of Plant Physiology Vol. 14B, eds. B. Parthier & D. Boulter, pp. 75-110. Berlin: Springer-Verlag.

Bryant, J.A., Jenns, S.M. & Francis, D. (1981). DNA polymerase activity and DNA synthesis in roots of pea (Pisum sativum L.) seedlings. Phytochemistry 20, 13-15.

Castroviejo, J., Tarrago-Litvak, L. & Litvak, S. (1975). Partial purification and characterization of two cytoplasmic DNA polymerases from ungerminated wheat. Nucleic Acids Research, 2, 2077-90.

Castroviejo, M., Tharaud, D., Tarrago-Litvak, L. & Litvak, S. (1979). Multiple deoxyribonucleic acid polymerases from quiescent wheat embryos. Purification and characterization of three enzymes from the soluble cytoplasm and one from purified mitochondria. Biochemical Journal 181, 183-91.

Chivers, H.J. (1984). DNA replication and the cell cycle in pea seedlings. Ph.D. thesis, University of Wales, U.K.

Chivers, H.J. & Bryant, J.A. (1983). Molecular weights of the major DNA polymerases in a higher plant, Pisum sativum L. (pea). Biochemical and Biophysical Research Communications 110, 632-9.

Cook, P.R. & Lang, J. (1984). The spatial organization of sequences involved in initiation and termination of eukaryotic DNA replication. Nucleic Acids Research 12, 1069-75.

D'Alesandro, M., Jaskot, R. & Dunham, V.L. (1980). Soluble and chromatin-bound DNA polymerases in developing soybean. Biochemical and Biophysical Research Communications 94, 233-9.

Dunham, V.L. & Bryant, J.A. (1981). DNA polymerases of turnip. (Brassica rapa). Biochemical Society Transactions 9, 230.

Dunham, V.L. & Cherry, J.H. (1973). Multiple DNA polymerase activity solubilized from higher plant chromatin. Biochemical and Biophysical Research Communications 54, 403-10.

Dunham, V.L. & Yunghans, W.N. (1977). Effects of nuclear proteins on the activity of soybean DNA polymerase. Biochemical and Biophysical Research Communications, 75, 987-94.

Durnford, J.M. & Champoux, J.J. (1978). The DNA untwisting enzyme from Saccharomyces cerevisiae. Journal of Biological Chemistry 253, 1086-19.

Dyan, W.S., Jendrisak, J.J., Hager, D.A. & Burgess, R.R. (1981). Purification and characterization of wheat germ DNA topoisomerase I (nicking-closing enzyme). Journal of Biological Chemistry 256, 5860-75.

Gardner, J.M. & Kado, C.I. (1976). High molecular weight deoxyribonucleic acid polymerase from crown gall tumour cells of periwinkle (Vinca rosea). Biochemistry 15, 688-96.

Goldberg, G.I., Collier, I. & Cassel, A. (1983). Specific DNA sequences associated with the nuclear matrix in synchronized mouse 3T3 cells. Proceedings of the National Academy of Sciences, USA. 80, 6887-91.

Graveline, J., Tarrago-Litvak, L., Castroviejo, M. & Litvak, S. (1984). DNA primase activity from wheat embryos. Plant Molecular Biology, 3, 207-216.

Harland, J., Jackson, J.F. & Yeoman, M.M. (1973). Changes in some enzymes involved in DNA synthesis following induction of division in cultured plant cells. Journal of Cell Science, 13, 121-38.

Herrick, G. & Alberts, B. (1976a). Purification and physical characterization of nucleic acid helix-unwinding proteins from calf thymus. Journal of Biological Chemistry 251, 2124-32.

Herrick, G. & Alberts, B. (1976b). Nucleic acid helix-coil transitions mediated by helix-unwinding proteins from calf thymus. Journal of Biological Chemistry 251, 2133-41.

Hotta, Y. & Stern, H. (1971). A DNA-binding protein in meiotic cells of Lilium. Developmental Biology 26, 87-99.

Hotta, Y. & Stern, H. (1978). DNA-unwinding protein from meiotic cells of Lilium. Biochemistry 17, 1872-80.

Hotta, Y. & Stern, H. (1979). The effect of dephosphorylation on the properties of a helix-destabilising protein from meiotic cells, and its partial reversal by a protein kinase. European Journal of Biochemistry 95, 31-8.

Hovemann, B. & Follmann, H. (1979). Deoxyribonucleotide synthesis and DNA polymerase activity in plant cells (Vicia faba and Glycine max). Biochimica et Biophysica Acta 561, 42-52.

Howell, S.H. & Stern, H. (1971). The appearance of DNA breakage and repair activities in the synchronous meiotic cycle of Lilium. Journal of Molecular Biology 55, 357-78.

Jakob, K.M. & Bovey, F. (1969). Early nucleic acid and protein synthesis and mitoses in the primary root tips of germinating Vicia faba. Experimental Cell Research 54, 118-26.

Jenns. S.M. & Bryant, J.A. (1978). Correlation between deoxyribonuclease activity and DNA replication in the embryonic axes of germinating peas (Pisum sativum L.). Planta 138, 99-103.

Kaguni, L.S., Rossignol, J., Conaway, R.C., Banks, G.R. & Lehman, I.R. (1983). Association of DNA primase with the β/γ subunits of DNA polymerase-α from Drosophila melanogaster embryos. Journal of Biological Chemistry 258, 9037-19.

Kende, H. (1971). The cytokinins. International Review of Cytology 31, 301-38.

Kessler, B. (1971). Isolation, purification and distribution of a DNA ligase from higher plants. Biochimica et Biophysica Acta 240, 496-505.

Koerner, T.J. & Meyer, R.R. (1983). A novel single-stranded DNA-binding protein from the Novikoff hepatoma which stimulates DNA polymerase-β. Journal of Biological Chemistry 258, 3126-33.

Kornberg, A. (1980). DNA replication. San Francisco: Freeman.

Kornberg, A. (1982). Supplement to DNA replication. San Francisco: Freeman.

Leffler, H.R., O'Brien, T.J., Glover, D.V. & Cherry, J.H. (1971). Enhanced deoxyribonucleic acid polymerase activity of chromatin from soybean hypocotyls treated with 2,4-dichlorophenoxyacetic acid. Plant Physiology 48, 43-5.

Misumi, M. & Weissbach, A. (1982). The isolation and characterization of DNA polymerase-α from spinach. Journal of Biological Chemistry 257, 2323-9.

Montiel, J.F., Norbury, C.J., Tuite, M.F., Donson, M.J., Mills, J.S., Kingsman, A.J. & Kingsman, S.M. (1984). Characterization of human chromosomal DNA sequences which replicate autonomously in Saccharomyces cerevisiae. Nucleic Acids Research 12, 1049-68.

Mory, Y.Y., Chen, D. & Sarid, S. (1975). De novo biosynthesis of deoxyribonucleic acid polymerase during wheat embryo germination. Plant Physiology 55, 437-42.

Nasmyth, K., Nurse, P. & Fraser, R.S.S. (1979). The effect of the cell mass on the cell cycle timing and duration of S-phase in fission yeast. Journal of Cell Science 39, 215-33.

Nurse, P. & Thuriaux, P. (1977). Controls over the timing of DNA replication during the cell cycle of fission yeast. Experimental Cell Research 107, 365-75.

Reddy, G.P.V. (1982). Catalytic function of thymidylate synthase is confined to S-phase due to its association with replitase. Biochemical and Biophysical Research Communications 109, 908-15.

Reddy, G.P.V. & Pardee, A.B. (1980). Multienzyme complex for metabolic channeling in mammalian DNA replication. Proceedings of the National Academy of Sciences, USA 77, 3312-6.

Robinson, N.E. & Bryant, J.A. (1975). Development of chromatin-bound and soluble DNA polymerase activities during germination of Pisum sativum L. Planta 127, 69-75.

Sala, F., Amileni, A.R., Parisi, B. & Spadari, S. (1980). A γ-like DNA polymerase in spinach chloroplasts. European Journal of Biochemistry 112, 211-7.

Sala, F., Galli, M.G., Levi, M., Burroni, D., Parisi, B., Pedrali-Noy, G. & Spadari, S. (1981). Functional roles of the plant α-like and γ-like DNA polymerases. FEBS Letters 124, 112-8.

Sala, F., Galli, M.G., Nielsen, E., Magnien, E., Devreux, M., Pedrali-Noy,G. & Spadari, S. (1983). Synchronization of nuclear DNA synthesis in cultured Daucus carota L. cells by aphidicolin. FEBS Letters 153, 204-8.

Sato, T., Watanabe, A. & Imaseki, H. (1976). Effect of ethylene on DNA synthesis in potato tuber discs. Plant and Cell Physiology 17, 1255-62.

Sawai, Y., Sugano, N. & Tsukada, J. (1978). Ribonuclease-H activity in cultured plant cells. Biochimica et Biophysica Acta 518, 181-5.

Schvartzman, J.B. & Van't Hof, J. (1982). In the higher plant Pisum sativum maturation of nascent DNA is blocked by cycloheximide, but only after 4-8 replicons are joined. Nucleic Acids Research 10, 6191-205.

Smith, C.A. & Okumoto, D.S. (1984). Nature of DNA repair synthesis resistant to inhibitors of polymerase-α human cells. Biochemistry 23, 1383-91.

Srivastava, B.I.S. (1974). A 7S DNA polymerase in the cytoplasmic fraction from higher plants. Life Science 14, 1947-54.

Stevens, C. & Bryant, J.A. (1978). Partial purification and characterization of the soluble DNA polymerase (polymerase-α) from seedlings of Pisum sativum L. Planta 138, 127-32.

Stewart, G.R. & Smith, H. (1972). Effects of abscisic acid on nucleic acid snythesis and induction of nitrate reductase in Lemna polyrhiza. Journal of Experimental Botany 23, 875-85.

Thomas, C.M., Bryant, J.A., Dunham, V.L. & Hull, R. (1983). Deoxyribonucleic acid polymerase activities in healthy and cauliflower-mosaic-virus infected turnip plants. Biochemical Society Transactions 11, 367-8.

Trewavas, A.J. (1976). Plant growth substances. In Molecular aspects of gene expression in plants. ed. J.A. Bryant pp. 249-98. London & New York: Academic Press.

Tsuchiya, E., Kimura, K., Miyakawa, T. & Fukui, S. (1984). Characteristic alteration in the nuclear DNA polymerase activity during the cell division cycle of Saccharomyces cerevisiae. Nucleic Acids Research 12, 3143-54.

Tymonko, J.M. & Dunham, V.L. (1977). Evidence for DNA polymerase-α and -β activity in sugar beet. Physiologia Plantarum 40, 27-30.

Van't Hof, J. (1976a). DNA fiber replication of chromosomes of pea root cells terminating S. Experimental Cell Research 99, 47-56.

Van't Hof, J. (1976b). Replicon size and rate of fork movement in early S of higher plant cells (Pisum sativum). Experimental Cell Research 103, 395-403.

Watanabe, A. & Imaseki, H. (1973). Induction of deoxyribonucleic acid synthesis in potato tuber tissue by cutting. Plant Physiology 51, 772-6.

Watanabe, A. & Imaseki, H. (1977). Enhancement of DNA polymerase activity in potato tuber slices. Plant and Cell Physiology 18, 849-58.

Weir, A.F. (1983). Characterization of endodeoxyribonuclease in pea (Pisum sativum L.) Ph.D. thesis, University of Wales, U.K.

Wyers, F., Sentenac, A. & Fromageot, P. (1973). Role of DNA-RNA hybrids in eukaryotes, Ribonuclease H in yeast. European Journal of Biochemistry 35, 270-81.

Yagura, T., Kozu, T., Seno, T., Saneyoshi, M., Hiraga, S. & Nagano, H. (1983). Novel form of DNA polymerase-α associated with DNA primase activity of vertebrates. Journal of Biological Chemistry 258, 13070-5.

DNA REPLICATION IN RELATION TO DNA *C* VALUES

D. Francis, A.D. Kidd and M.D. Bennett

INTRODUCTION

DNA replication in eukaryotes consists of a series of events modulated by a multi-enzyme complex (Bryant, 1982; Dunham & Bryant, this volume). How it is regulated is unknown. However, it is known that the time taken to replicate the nuclear genome varies greatly between organisms, even at constant temperature. For example, at 23° the duration of DNA synthetic (S)-phase of the mitotic cell cycle was only 2.8 h in *Arabidopsis thaliana*, while in *Allium tuberosum* it was 11.8 h (Van't Hof, 1965).

One factor which may affect S-phase duration is the total amount of DNA to be replicated. Indeed, the duration of S-phase in unrelated diploid plants was shown to be a function of the amount of nuclear DNA (Van't Hof & Sparrow, 1963; Van't Hof, 1965; Evans & Rees, 1971; Evans, Rees, Snell & Sun, 1972). Thus, S-phase duration tends to increase with increasing DNA *C* value (defined as the amount of DNA in the unreplicated haploid genome). Such an increase in S-phase duration could be caused by three independent mechanisms. First, the rate of replication could decrease. Second, the number of replicons per unit length of DNA could decrease. Third, there could be an increase in the number of replicon families sequentially involved in DNA synthesis. Moreover, these mechanisms are not necessarily mutually exclusive.

How differences in S-phase duration between species with contrasting DNA *C* values are determined, is largely unknown. Consequently, work which increases our knowledge of these aspects of DNA synthesis in species with contrasting DNA *C* values is desirable.

The increase in duration of S-phase accompanying increases in *C* value, found in diploid angiosperms, reflects neither an increase in the overall genetic complexity of the genome, nor a change in the chemical process of replication. Rather, it appears to be correlated primarily with a physical character, the mass of DNA per basic genome. The term 'nucleotype' was suggested (Bennett, 1971) to describe those conditions of the nucleus that affect the phenotype independently of the informational content of the nuclear DNA. Evidence has accumulated showing that in higher plants strong correlations exist between nuclear DNA *C* values and a wide range of cellular and whole plant characters including minimum generation time and geographical distribution. It was suggested that many of these correlations are largely nucleotypic effects (Bennett, 1972, 1976). Our aims in this chapter are (1) to review several nucleotypic correlations in relation to DNA synthesis, and (2) to collate rates of DNA synthesis and estimates of replicon size in 19 angiosperm species with contrasting DNA *C* values. The latter include all the published data of which we are aware, together with our unpublished results.

DNA C VALUE PARADOX

A discussion of nuclear DNA amounts would be incomplete without mentioning the *C* value paradox. However, we do not intend to retrace ground adequately covered elsewhere (e.g. Cavalier-Smith, 1978). Eukaryotes display a 40,000-fold range of nuclear DNA *C* values from 0.005 pg in yeast (Sokurova, 1973) to 200 pg in the dinoflagellate *Gonyaulaux* (Holm-Hansen, 1969). Observations in lower and higher plants (Holm-Hansen, 1969; Sparrow, Price & Underbrink, 1972; Bennett & Smith, 1976) show that large interspecific increases in DNA *C* value both within a genus, and even within major groups of eukaryotes, are often not accompanied by any increase in the amount, type and control of gene expression. For example, the angiosperms exhibit at least a 600-fold range of DNA *C* values (Bennett & Smith, 1976). Within this range, the dicot genus *Vicia* exhibits a seven-fold variation from 2.0 pg in *V. villosa* to 14.4 pg in *V. faba* (Maher & Fox, 1973). Similarly, the monocot genus *Allium* shows a five-fold variation in DNA *C* values (Jones & Rees, 1968). Despite this variation in genome size, all angiosperm species have similar levels of genic complexity.

As no explanation for such interspecific variation in genome size was forthcoming, Thomas (1971) described this problem as the "*C* value paradox" (see also Lewin, 1974, 1980). The problem was increased with the discovery that the genomes of many eukaryotes contain large numbers of repeated DNA sequences many of which are non-genic, i.e. sequences which do not code for any RNA molecule. Indeed, non-genic sequences can account for 90% of the total nuclear DNA (Flavell, 1980). Although no function can be ascribed to non-genic repetitive DNA sequences, clearly they do not normally interfere with chromosome functions. Moreover, Flavell (1980) noted that constitutive heterochromatin, consisting of repetitive nucleotide sequences, may determine the availability of chromosomal sites as origins of replication.

NUCLEOTYPIC EFFECTS

While the full significance of the variation in DNA C values is probably still unknown, our knowledge of the consequences of variation in genome size on plant phenotypic characters shows clearly that large changes in genome size are far from neutral in their effects on organisms. As noted above, the angiosperms exhibit at least a 600-fold variation in DNA C values between species, and this variation is strikingly correlated with the duration of developmental processes at the cellular level, provided only corresponding cells from species with a single ploidy level, grown under constant environmental conditions, are compared. For example, Figure 1 illustrates a strong positive correlation between nuclear DNA mass and the duration of the mitotic cycle in the root meristem in a range of diploid monocots ($r = 0.99$; $P<0.001$). A similar correlation ($r = 0.96$; $P<0.001$) has been shown between DNA C value and the duration of meiosis in diploid monocots (Fig. 2).

It is important to note that these close correlations exist between mitotic and meiotic duration and DNA C value, even though unrelated diploid species, subjected to widely different selection pressures, are compared (Bennett, 1972). However, if samples of dissimilar cells or species are compared, these close correlations disappear. For example, when closely related cereal species with _different_ ploidy levels were compared, the duration of meiosis decreased with increasing nuclear DNA C value (Table 1; Bennett & Smith, 1972). Similarly, the mitotic cell cycle time did not increase with increasing nuclear DNA C value when related cereals with different ploidy levels were compared. For example, although hexaploid triticale (21.2 pg) has a higher DNA C value than either of its parental species (8.8 and 12.3 pg, respectively) the cell cycle in root meristem cells at 20° was not significantly longer in hexaploid triticale (12 h) than in either of its parent species, tetraploid *Triticum turgidum* (13.8 h) and diploid *Secale cereale* (cv. Prolific) (11.5 h) (Kaltsikes, 1971). Thus the durations of both meiosis and the mitotic cell cycle do not invariably increase in allopolyploids compared with their related diploids, despite the increased nuclear DNA content of the former. Clearly, there is not a

Fig. 1. The relationship between duration of the cell cycle (h) and DNA C value (pg), in the root meristem of seven diploid, monocot species at 20° (adapted from Bennett, 1972).

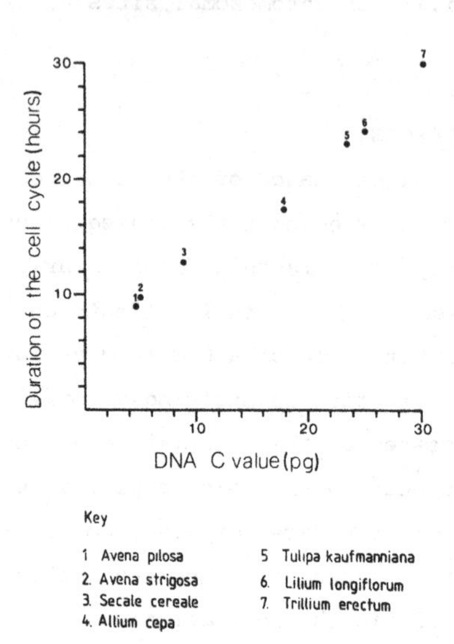

Fig. 2. The relationship between duration of meiosis (h) and DNA C value (pg), for 10 diploid, monocot species at 20° (adapted from Bennett, 1977a).

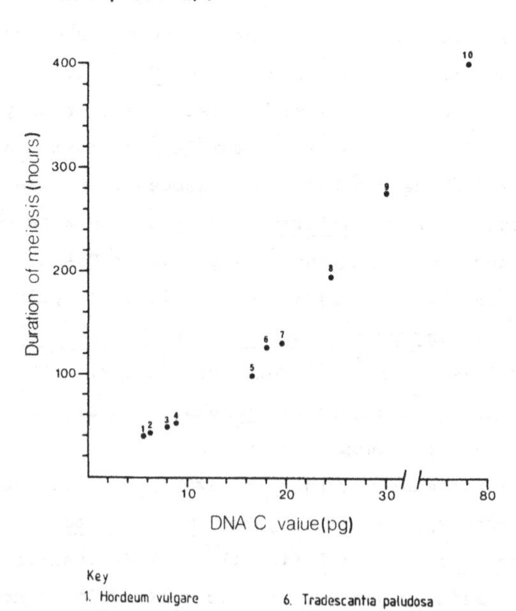

simple and absolute correlation between nuclear DNA C value and the rates of cell development at meiosis or mitosis. Other factors besides nuclear DNA C value have important effects on these characters. Nevertheless, when estimates of the duration of meiosis for species at each ploidy level are plotted separately, meiosis is longer in the species with the higher C value within each ploidy level (e.g. see Table 1). Indeed, nuclear DNA content and meiotic duration were positively correlated at each ploidy level, but the slope of the relationship increased with increasing ploidy level (Bennett, 1977a). Which other characters are causally responsible for the differences between ploidy levels is uncertain, but clearly there is an effect of DNA amount on the rate of meiotic development within each ploidy level.

Nucleotypic effects are not confined to the cellular level, but extend to characters of tissues, organs and even whole organisms. For example, DNA C value is positively correlated with the minimum generation time i.e. the minimum time taken from germination to the first production

Table 1. The duration of meiosis (h) at 20° and DNA C value in 10 cereal species

Ploidy level	Genotype	C value (pg)	Duration of meiosis (h)
Diploid	*Hordeum vulgare*	5.5	39.4
	Triticum monococcum	6.2	42.0
	Secale cereale	8.8	51.2
Tetraploid	*Hordeum vulgare*	11.0	31.0
	Triticum dicoccum	12.0	30.0
	Secale cereale (cv. Svälof)	17.2	38.0
Hexaploid	*Triticum aestivum* (cv. Chinese Spring)	17.3	24.0
	Triticale (Rosner)	21.2	34.0
Octoploid	*Triticale* (Chinese Spring x King II)	26.0	20.8
	Triticale (Chinese Spring x Petkus Spring)	26.0	22.0

of a mature seed, in herbaceous angiosperms (Bennett, 1972). All species with very short minimum generation times (e.g. all ephemerals) have C values low in the range known for angiosperms. Conversely, all species known to have very high C values have long minimum generation times, and hence are perennials. Thus, the life cycle of just a few weeks exhibited by ephemerals requires a low DNA C value as well as the necessary genic controls. Species with low DNA C values have the 'option' to express a wide range of life cycle types, and hence may be ephemerals, annuals or perennials, subject to genic control. In contrast, species with very high DNA C values have no option to express an ephemeral or an annual life cycle. Their nucleotype invariably dictates a long minimum generation time making them all obligate perennials.

Typical examples of species with contrasting nucleotypes are *Arabidopsis thaliana* and *Trillium erectum*. *A. thaliana* has a DNA C value of 0.2 pg; its root meristem cell cycle time, at 23°, is about 8 h (Van't Hof, Kuniyuki & Bjerknes, 1978), and its minimum generation time is about six weeks. On the other hand, *T. erectum*, which has a C value of 30 pg, has a root meristem cell cycle, at 20°, of at least 29 h. It cannot complete a life cycle within a year, and so it is an obligate perennial. These contrasting types show clearly that the maximum rate of development at the cellular level can be related to that of the whole organism.

A significant positive correlation has also been demonstrated between DNA amount per diploid genome and the latitude at which man chooses to cultivate his major crops, namely, cereal grains, pasture grasses and pulses (Bennett, 1976). This nucleotypic cline for DNA amount per diploid genome on latitude was seen particularly clearly in comparisons of 20 cereal grain species. For example, of these species only five were cultivated at latitudes higher than 60°N and these had a mean C value per diploid genome of 5.6 pg. This was significantly higher than the corresponding mean of 1.8 pg, for five other species whose cultivation is confined to latitudes below 47°N. The cline appears to be a natural phenomenon which is modified, and perhaps exaggerated in agriculture.

Correlations like those described above show that gross variation in DNA amount between angiosperm species can have adaptive significance. Nuclear DNA C value can affect cell cycle activity and growth habit, and hence may well play a role in determining both the optimum environment and the geographic range for both crop and non-crop species (Bennett, 1976). However, the question remains as to how these nucleotypic effects are determined and linked. The rate at which nuclear DNA is doubled during S-phase of the cell cycle may be partly responsible, since the greater the amount of nuclear DNA the longer the time required to replicate it (Van't Hof & Sparrow, 1963). This would contribute to a lengthened cell cycle which, it has been suggested, could eventually affect minimum generation time and perhaps the geographic distribution of species. However the relationship, if any, between DNA C value and the basic aspects of DNA synthesis have yet to be fully explored. In order to investigate any such relationships it would be necessary to determine the number of replicons per nuclear genome, and rates of nuclear DNA replication in replicons, of many species. The angiosperms with their wide range of DNA C values (Bennett & Smith, 1976) are clearly excellent materials for an investigation of this problem.

DNA REPLICATION AND FIBRE AUTORADIOGRAPHY

DNA synthesis in eukaryotes occurs during S-phase of the cell cycle and involves multiple replicons spaced along the DNA molecule (see Waterborg & Shall, this volume). Indeed, DNA replication is characterised by large numbers of these replicons per nucleus. However, the number of replicons per nucleus may vary greatly between species. For example, replicating nuclei of cultured cells of *Xenopus laevis* (C value = 3 pg) contain about 30,000 replicons (Callan, 1972), while the mean replicon size of 175 μm (i.e. about 500,000 base pairs) in nuclei of primary cultured cells of *Triturus cristatus-carnifex* (C value = 30 pg), is consistent with only about 1,060 replicons per nucleus (Callan, 1972). In comparison, root meristem nuclei of *Pisum sativum* cv. Alaska (C value = 4.6 pg) and *Secale cereale* cv Petkus Spring (C value = 8.8 pg) have about 180,000 and 250,000 replicons, respectively (Van't Hof, 1976; D. Francis, unpublished data). Moreover, replicon size can also differ between stages of development within an organism (e.g. Blumenthal, Kriegstein & Hogness, 1974). Thus, for inter-specific comparisons, it is important to compare replicon sizes for corresponding tissues at a given stage of development.

Initiation of DNA synthesis occurs at many points throughout the chromosome in groups of replicons commonly referred to as replicon families (De Pamphilis & Wassarman, 1980). Not all replicons replicate simultaneously, and the pattern of replicon activation during S-phase is probably species-specific. Indeed, the way in which replicon families function during different stages of S-phase may contribute to nucleotypic effects.

Each replicon has an origin and two diverging forks (Fig. 3).

Replication occurs bidirectionally as the double helix is denatured and unwound at the origin. Each diverging fork replicates, and new nucleotides are added in the 5' to 3' direction of the sugar-phosphate backbone of the molecule (Kornberg, 1980) so that nascent DNA is synthesised in a semi-conservative manner. The leading 5' to 3' strands moving towards the termini give the replicon its bidirectionality (De Pamphilis & Wassarman, 1980). The distance between one origin of replication, and an origin of an adjacent replicon will effectively measure the size of a replicon.

Both replicon size, and the rate at which a single fork replicates, can be determined using the technique of DNA fibre autoradiography. This technique, originally developed on Chinese Hamster cells (Huberman & Riggs, 1966) was successfully adapted for use with plant cells by Van't Hof (1975). Plant meristems are exposed to high specific activity (70-90 Ci/mmol) tritiated (methyl-^{3}H)-thymidine (^{3}H-TdR) for various durations, and labelled nuclei are isolated on microscope slides. The labelled chromatin is then treated with either trypsin, or a pronase, and the released DNA fibres are spread across the slide and prepared as DNA fibre autoradiographs.

After development, silver grain arrays are seen running as distinct tracks across the slide (Fig. 4). Replicon size can be determined by estimating the distance from the gap between one symmetrical pair of grain arrays to the corresponding gap in an adjacent pair. However, this

Fig. 3. Diagrammatic representation of a replicon. The double helix unwinds from the replicon origin (O) exposing complementary strands of the parental DNA molecule. Replication of daughter strands (----) is in the 5' to 3' direction. Replication forks proceed towards the termini (T) as the parent molecule unwinds.

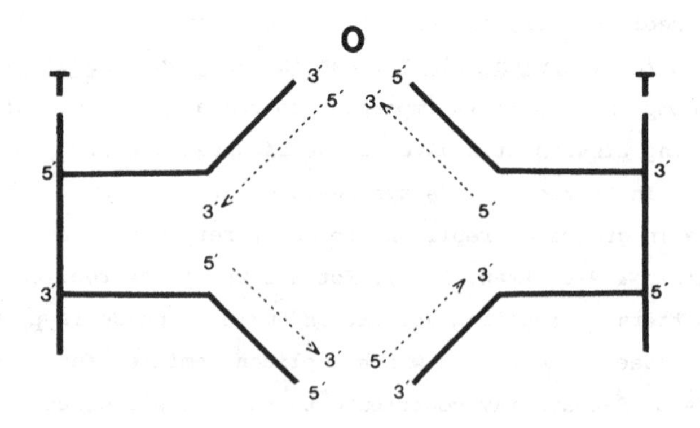

method assumes that each gap represents a putative replicon origin. Replicons which began to synthesise during a pulse with ^{3}H-TdR would also be represented by single tracks of silver grains. Consequently another procedure was developed (Huberman & Tsai, 1973). Meristems are exposed to high-, followed by low-specific activity (2Ci/mmol) ^{3}H-TdR. Replicons initiating DNA synthesis during exposure to high specific activity ^{3}H-TdR are characterised by a silver grain array with a dense central region flanked on either side by less dense regions. Replicons active before exposure to ^{3}H-TdR have gaps between the dense concentration of silver grains. On these autoradiographs replicon size is determined by measuring centre to centre distances between adjacent silver grain arrays.

Rates of replication per single replicon fork can be obtained

Fig. 4. Autoradiograph of isolated DNA fibres of *Zea mays* root tips pulse labelled with high specific activity ^{3}H-TdR at 20° for 120 min. The arrows point toward the location of replicon origins. Bar scale represents 20 μm.

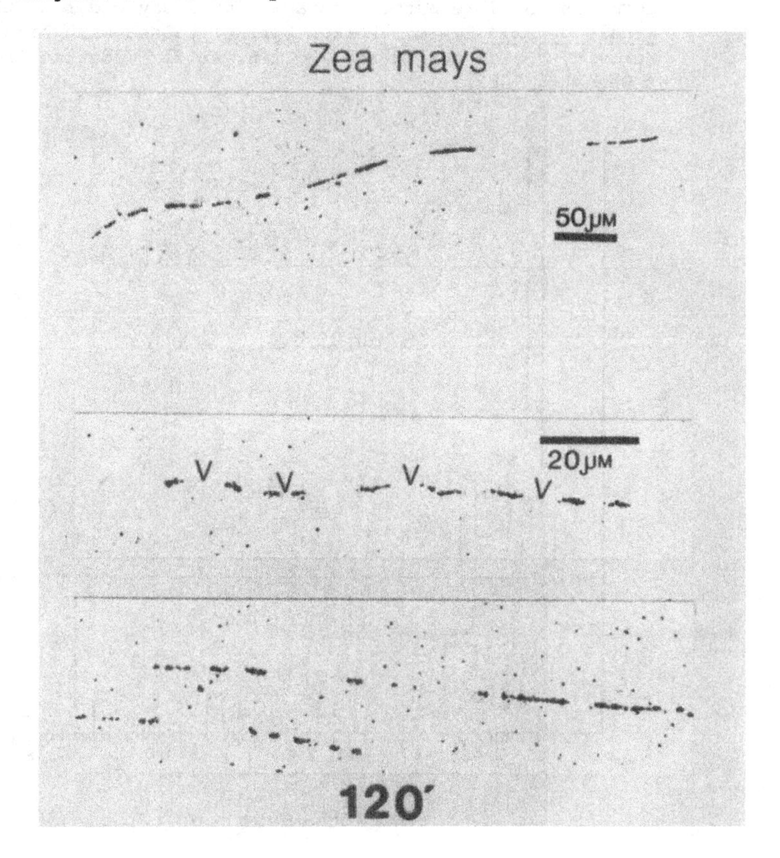

by establishing the relationship between mean fibre length from tandem arrays of silver grains, and the duration of a single exposure to high specific activity ^{3}H-TdR.

REPLICON SIZE AND RATES OF DNA SYNTHESIS IN ANGIOSPERMS

Estimates of replicon size from fibre autoradiographs normally display modal distributions for a given pulse with ^{3}H-TdR (Fig. 5). Generally, the shorter the exposure to ^{3}H-TdR the more pronounced a particular modal size becomes. As pulse time is prolonged, smaller secondary and tertiary peaks should appear representing arrays whose lengths are multiples of the basic modal peak and hence, multiples of the length of a single replicon. For example, exposure of four-day-old root meristems of *Secale cereale* cv. Petkus Spring to ^{3}H-TdR for 15 min at 23°C gave a modal size in the class

Fig. 5. Frequency distribution of mid-point to mid-point distances (µm) between adjacent tandem pairs of labelled DNA segments on fibre autoradiographs for 4 day old root tips of *Secale cereale* L. (cv. Petkus Spring) pulsed with high specific activity ^{3}H-TdR for 15 and 120 min. at 23° (adapted from Francis & Bennett, 1982).

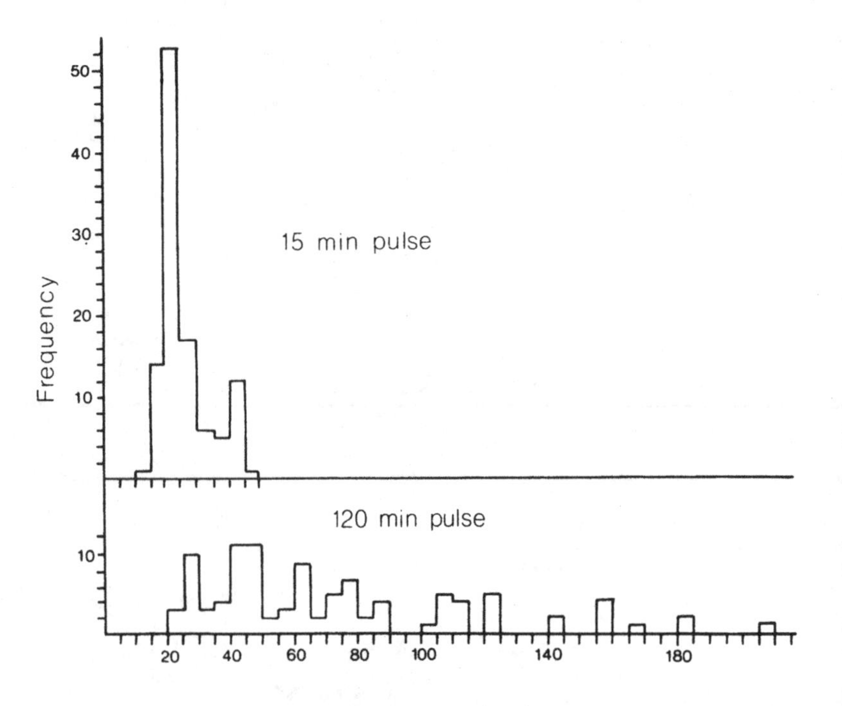

range 20-25 μm, but extending the exposure to 120 min gave a wider range of values for lengths of continuous arrays of silver grains which also contained other peaks for multiples of the basic 20-25 μm modal class size, i.e. 40-50 μm and 50-65 μm (Fig. 5). The larger class sizes for longer pulses of ^{3}H-TdR are consistent with fusion of forks from adjacent replicons, or measurements which span adjacent tandem arrays for replicons which either were active before the pulse or straddle other replicons which were quiescent during the pulse. Estimates of replicon size also typically include measurements smaller than the modal size. Thus, fibre autoradiography gives a measure of major class size, rather than demonstrating the existence of a replicon of constant definitive size, distributed everywhere throughout the chromosomes (see also Waterborg & Shall, this volume).

To our knowledge, fibre autoradiography has been applied to 19 angiosperm species. The results for four to five day-old root meristem cells are presented in Table 2 and Figure 6. It should be noted that while some estimates of replicon size were made at 20°, others were made at 23°. Van't Hof, Bjerknes & Clinton (1978) detected no difference between replicon size in Helianthus annuus cv. Russian Mammoth grown at 20 or 23°. Similarly, the modal replicon size was apparently unaltered in our experiments (A.D. Kidd, D. Francis & M.D. Bennett, unpublished data) despite variation in temperature. In Secale cereale it was within the class range 20-25 μm both at 20 and 23°, while in Oryza sativa it was in the class range 15-20 μm both at 20 and 27°. Thus, pooling estimates of replicon size obtained at 20 or 23°, as in Table 2, seems justified.

A conserved replicon size of 20-40 μm was suggested for all eukaryotes (Van't Hof, Bjerknes & Clinton, 1979). Most estimates for dicots are in, or close to, this range (Table 2, Fig. 6). However, only two of the corresponding estimates for eleven monocot species, which range from 5 to 21 μm, fall in the range 20-40 μm. Indeed, the mean replicon size for eight dicot species (22.2 μm) is significantly higher ($P < 0.002$) than for eleven monocots (14.4 μm). Similarly, the mean replicon size for seven diploid dicots (21.8 μm) is significantly higher ($P = 0.02-0.01$) than the mean for eight diploid monocots (16.7 μm).

Estimates for three polyploid monocots are all below the mean for diploid monocots, and strikingly so for two allohexaploids, Triticum aestivum and triticale, where the mean replicon size was only about 5 μm (Table 2). Clearly, further analyses of many more species will be required to discover whether or not a replicon size of 20-40 μm is truly representative of eukaryotes. To this end, further studies should include more angiosperms, both because of their large interspecific range of genome sizes, and because the above comparisons indicate that mean replicon size may tend to differ between monocots and dicots, and between diploids and polyploids.

Also relevant to the question of whether eukaryotes have a conserved replicon size is the possibility of intraspecific variation in replicon size. Variation in the S-phase duration clearly shows that nuclei from different tissues within an angiosperm can synthesize their DNA at different rates. For example, in Triticum aestivum, S-phase duration was no more than 4 h in young endosperm nuclei, but about 10 h at premeiotic interphase (Bennett, Rao, Smith & Bayliss, 1973). Whether such variation in S-phase duration in plants involves variation in the rate of fork movement,

Fig. 6. The relationship between average replicon size (µm) and DNA C value for cells in the root meristem of diploid dicots (▲); tetraploid dicots (Δ); diploid monocots (●) and polyploid monocots (O) at 20 or 23° (see Table 2).

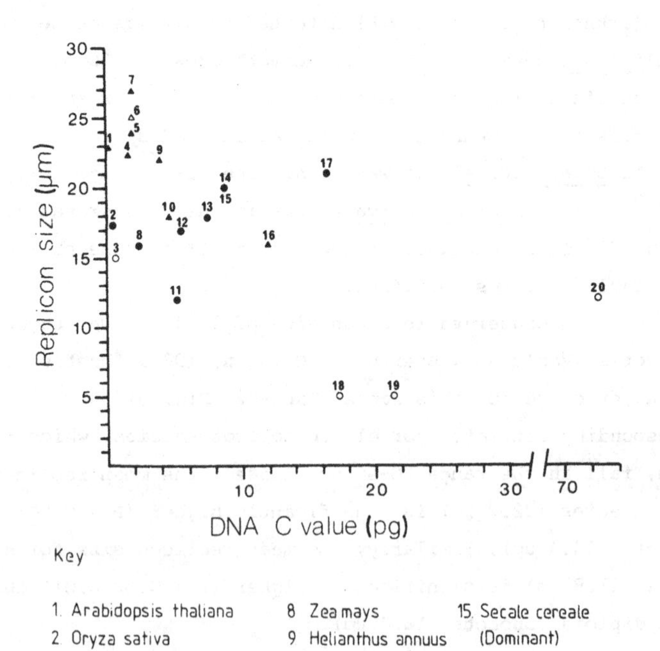

Table 2. Chromosome number per genome, DNA C value (pg), average rate of DNA replication (μm/h), replicon size (μm), Rs (h), Ds (h), and Rs:Ds ratio in 4 day old root meristems of 19 diploid, tetraploid and hexaploid angiosperm species at 20 or 23°

	Chromosome no. (2n)	C value (pg)	Rate of Replication (μm/h)	Average Replicon size (μm)	Rs (h)	Ds (h)	Rs:Ds ratio	Temp.	Dicot (D) Monocot (M)	Ploidy
1. *Arabidopsis thaliana*[a]	10	0.2	5.8	23	2.00	2.8	0.71	23°	D	2x
2. *Oryza sativa*[b]	24	0.6	6.0	17.5	1.50	4.0[h]	0.37	20°	M	2x
3. *Eragrostis tef*[b]	40	0.7	8.2	15.0	0.91	-	-	20°	M	4x
4. *Phaseolus coccineus*[c]	22	1.7	9.6	22.5	1.20	-	-	23°	D	2x
5. *Lycopersicon esculentum*[a]	24	1.9	8.0	24.0	1.50	4.3	0.34	23°	D	2x
6. *Glycine max*[a]	40	1.9	6.8	25.0	1.80	5.0	0.36	23°	D	4x
7. *Haplopappus gracilis*[a]	4	2.0	6.9	27.0	2.00	4.0	0.50	23°	D	2x
8. *Zea mays* (cv. Seneca 60)[b]	20	2.4	3.0	16.0	2.70	8.0[e]	0.34	20°	M	2x
9. *Helianthus annuus*[a]	34	4.0	10.0	22.0	1.10	5.0	0.22	23°	D	2x
10. *Pisum sativum*[a]	14	4.6	9.0	18.0	1.00	5.0	0.20	23°	D	2x
11. *Aegilops squarrosa*[b]	14	5.1	3.2	12.0	1.88	6.2[g]	0.30	20°	M	2x
12. *Hordeum vulgare*[b]	14	5.5	4.1	17.0	2.10	3.8[f]	0.55	20°	M	2x
13. *Secale africanum*[b]	14	7.4	1.9	18.0	4.70	-	-	20°	M	2x
14. *Secale cereale*[d] (cv. Petkus Spring)	14	8.8	12.1	20.0	0.83	-	-	23°	M	2x
15. *Secale cereale*[b] (cv. Dominant)	14	8.8	7.3	20.0	1.37	6.6	0.21	20°	M	2x
16. *Vicia faba*[a]	12	11.9	9.0	16.0	0.90	7.8	0.11	20°	D	2x
17. *Allium cepa*[a]	16	16.5	7.5	21.0	1.40	10.9	0.12	23°	M	2x
18. *Triticum aestivum*[b] (cv. Chinese Spring)	42	17.3	1.6	5.0	1.56	4.6	0.34	20°	M	6x
19. *Triticale*[b] T7	42	21.2	1.0	5.0	2.50	4.7	0.53	20°	M	6x
20. *Fritillaria imperialis*[b]	24	72.5	4.7	12.0	1.28	-	-	20°	M	2x

a. Van't Hof & Bjerknes (1981)
b. Kidd, Francis & Bennett (unpublished)
c. Francis (unpublished)
d. Francis & Bennett (1982)
e. Verma (1980)
f. Bennett & Finch (1972)
g. Davies & Rees (1975)
h. Kowyama (1983)

Fig. 7. The relationship between rate of replication per single replicon fork (μm/h) and DNA C value for cells in the root meristem of diploid (▲) and tetraploid (Δ) dicots, and diploid (●) and polyploid (O) monocots at (a) 23° or (b) 20°. (see Table 2).

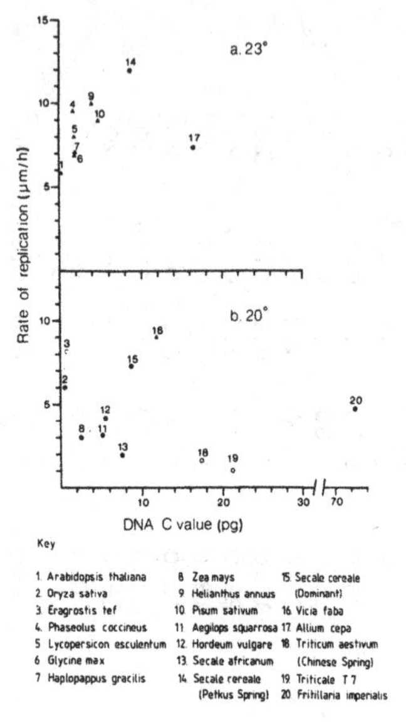

Fig. 8. The relationship between Rs:Ds ratios and DNA C value for cells of the root meristem of diploid (▲) and tetraploid (Δ) dicots, and diploid (●) and polyploid (O) monocots (see Table 2).

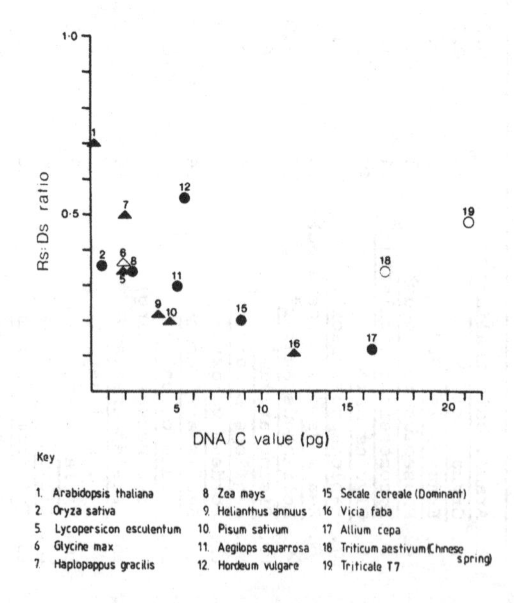

or changes in replicon size, or both, is unknown. However, studies of animal cells showed that intraspecific variation in S-phase duration, e.g. in Triturus (Callan, 1972), can involve large changes in replicon size; faster DNA synthesis involved shorter replicons.

Table 2 lists estimates for the rate of fork movement in four to five day old root meristem cells of 19 angiosperm species, obtained at either 20 or 23°. However, rates were estimated at both temperatures in two of these species, Helianthus annuus and Secale cereale. In H. annuus cv. Russian Mammoth, the mean rate of fork movement at 20 and 23° was 8 and 10 μm/h, respectively (Van't Hof et al., 1978). In S. cereale cv. Dominant, which is a winter variety, fork movement at 20° was 7 μm/h, while in the variety Petkus Spring, which is a spring type, fork movement at 23° was 12 μm/h. It is unclear whether the difference in rate of fork movement in S. cereale is due to temperature, or to inter-varietal genetic variation. However, as there was a 1.25-fold increase in rate of fork movement in H. annuus, and a 1.7-fold increase in S. cereale, between 20 and 23°, rate of fork movement is plotted against DNA C value separately for estimates made at 20° (Fig. 7b), and at 23° (Fig. 7a). It should be noted that this tended to separate most estimates for dicots (Fig. 7a) from most estimates for monocots (Fig. 7b).

Analysis of the results plotted in Figure 7a showed no significant relationship between rate of fork movement and DNA C value either using a linear regression analysis ($r = 0.29$; $P > 0.1$) or a Spearman Rank correlation test ($r_s = 0.38$; $P > 0.1$). In the latter, no assumptions were made about linearity. Similarly, no significant relationship was found for results plotted in Figure 7b using either statistical test ($r = -0.12$; $r_s = -0.44$; $P > 0.1$). It should be noted, however, that these comparisons may be confounded by pooling species with different S-phase durations, and different ploidy levels. While no firm conclusion can be drawn until more results are obtained, the available evidence indicates that close relationships between rate of fork extension and DNA C value seem unlikely, either for angiosperms in general, or separately for monocots and dicots.

In an attempt to overcome the confounding effects due to interspecific variation in S-phase duration, and to temperature differences, it seemed meaningful to express the time required for a replicon to replicate its allotted DNA (Rs) in relation to the duration of S-phase (Ds: referred to by Van't Hof & Bjerknes (1981) as Ts). Rs is calculated by dividing average replicon size by twice the rate of fork extension (Van't Hof & Bjerknes, 1981). Most measurements of S-phase, rates of replication and replicon size in a particular species in Table 2, were for root meristems at the same stage of development, from the same cultured variety. However, where data on S-phase were lacking published values for the species were used. For example, the value of 8 h for S-phase duration in maize was published by Verma (1980) for four day-old root tips of *Zea mays* cv. Seneca 60 at 20° (Table 2).

An Rs:Ds ratio of 1 is expected when the genome comprises one replicon, as in most prokaryotes, or several replicons which are all synchronously activated at the beginning of S-phase. In eukaryotes the Rs: Ds ratio would approach unity only if all replicons are of equal size, have similar fork rates and, importantly, initiate DNA synthesis simultaneously. Conversely the ratio will decrease when replicons are activated and especially if the time interval between activation of different replicon families becomes protracted.

Rs:Ds ratios have been calculated for 15 species, i.e. all those for which the time to replicate a replicon and the duration of S-phase have been estimated. These varied from 0.71 for *Arabidopsis thaliana* to 0.11 for *Vicia faba* (Table 2). As noted by Van't Hof, Kuniyuki and Bjerknes (1978), the Rs:Ds ratio of 0.71 for *Arabidopsis thaliana* is consistent with a low DNA *C* value (0.2 pg), and only two replicon families which initiate DNA synthesis with an interval of 36 min between them. Conversely, a low ratio of 0.12, like that obtained for *Allium cepa*, is consistent with an asynchronous activation of several replicon families in a species with a high DNA *C* value (16.5 pg) and a long S-phase (10.9 h at 23°; Van't Hof & Bjerknes, 1981).

Figure 8 is a plot of Rs:Ds ratios for all 15 species against their DNA *C* values. Linear regression analysis revealed a highly significant negative correlation ($r = -0.69$; $P = 0.02-0.01$) between these two characters for 12 diploids; the Spearman Rank correlation was also significant ($r_s = -0.73$; $P = 0.01-0.001$). Thus for diploids, the higher the DNA *C* value, the lower is the Rs:Ds ratio. However, when the results for all species, including polyploids, are analysed by either test, the correlations were non-significant ($r = -0.37$; $r_s = -0.28$; $P > 0.1$). This suggests that polyploids may not conform to the same relationship as diploids, as previously noted for other characters e.g. meiotic duration and DNA *C* value. Further analysis showed that the points for the two hexaploid monocots *Triticum* aestivum and triticale departed significantly from the linear regression line for diploids, unlike that for the tetraploid dicot Glycine max.

It is therefore interesting to compare DNA synthesis in these two allohexaploids with that in two diploids Allium cepa and Vicia faba with similar DNA C values. For example, the DNA C values for diploid A. cepa and hexaploid T. aestivum are 16.5 and 17.5 pg, respectively. The somewhat higher Rs:Ds ratios for T. aestivum (0.34) and triticale (0.53), compared with those for A. cepa (0.12) and Vicia faba (0.11), are consistent with a more synchronous activation of replicons in the polyploids than in the diploids. The substantial differences between these two allohexaploids and the two diploids are the very low rates of single fork extension and the much smaller replicon size in the former than in the latter (Fig. 6-7; Table 2).

It seems worthwhile to comment on DNA synthesis in triticale, an allohexaploid whose nuclei contain 28 wheat chromosomes and 14 rye (S. cereale) chromosomes. Fibre autoradiography for this species has failed to reveal a pattern of fibres resembling those seen for S. cereale. Instead, patterns of fibres from triticale were all remarkably similar to those from hexaploid Triticum aestivum. Thus, the average replicon size in triticale (5 μm) was identical to that in hexaploid T. aestivum, but much smaller than that in S. cereale (20 μm) (Table 2). Similarly, the rate of single fork extension in triticale (1.0 μm/h) was much more similar to that of T. aestivum (1.6 μm/h) than those of two S. cereale cultivars (7.8 and 12.1 μm/h, respectively). Also, the synchrony of DNA synthesis signified by the Rs:Ds ratio for triticale (0.53) was more like that of hexaploid T. aestivum (0.34) than that of diploid S. cereale (0.21). This indicates a more synchronous activation of replicons in rye chromosomes in hexaploid triticale than in the diploid parental species, S. cereale. Thus, it appears as if genes in the triticale nucleus, presumably in the wheat chromosomes, are able to regulate replicon size, rate of fork movement, and even synchrony of replication in the entire triticale genome, and to impose on the rye chromosomes behaviour quite unlike that found in corresponding cells of the diploid species S. cereale, but more like that of allopolyploid wheat. Clearly, data for the tetraploid wheat parent of this hexaploid triticale are needed in order to fully substantiate this interpretation, but the remarkable similarity between the various aspects of DNA synthesis in hexaploid triticale and hexaploid T. aestivum is entirely consistent with this hypothesis. Meanwhile it is worth noting that the imposition of alien DNA synthesis behaviour on rye chromosomes in triticale, e.g. the imposition of a slower rate of fork movement, might sometimes result in incomplete ligation of, or failure to replicate, some rye replicon families. If so, this might be a cause of incomplete DNA synthesis in heterochromatic telomeres of rye chromosomes, which has been suggested as one cause of aberrant endosperm nuclei and of grain shrivelling in triticales (Bennett, 1977b, 1980).

CONCLUSIONS

Comparisons of DNA C value with either estimates of replicon size, or rate of single fork extension for 19 angiosperm species with contrasting DNA C values, revealed no significant relationships ($P > 0.1$). Thus, the positive correlations noted between DNA C value and S-phase (see Introduction for references) are probably not explained on the basis of either larger replicons or slower rates of DNA synthesis, with increasing DNA C value.

A character which may be more meaningful in explaining nucleotypic effects is the Rs:Ds ratio which expresses the time taken to replicate the DNA in a modal replicon in relation to duration of S-phase. This ratio is largely, therefore, a function of the overall asynchrony of replication of all replicons in the genome, both within and between replicon families.

Analyses in the present work showed that for a sample of 12 diploid angiosperm species the Rs:Ds ratio was inversely correlated with DNA C value ($r = -0.69$), and significantly so ($P = 0.02$-0.01). Thus, as DNA C value in diploids is increased, the more asynchronous is the activation of replicons and replicon families, and the greater the time interval between their activation, the longer S-phase becomes.

Greater asynchrony of replicon activation with increasing DNA C value per basic genome might depend on changes in nuclear architecture independent of whether the additional DNA sequences are genic or non-genic, unique or repeated, e.g. on changes in the intranuclear distances between controlling genes and controlled chromatin segments. On the other hand, greater asynchrony of activation or replicons may be influenced primarily by which type of DNA sequence is increased. Increases in DNA C values often involve increases in the total amount of repetitive sequences per basic genome (Flavell, Bennett, Smith & Smith, 1974; Nagl, 1982). Perhaps such increases are particularly associated with increasing asynchrony of activation of replicons within and between replicon families. Repetitive sequences which are devoid of genic information might nevertheless modulate the expression of genes which control the pattern of DNA synthesis, such that the greater the proportion of repetitive non-genic DNA, the greater the effect (see Nagl & Scherthan, this volume).

Clearly there is no absolute relationship between increased DNA C value and the degree of synchrony of replicon activation within the nuclear genome. Thus, the present results for triticale (Rs:Ds = 0.53) suggest that replicon activation is more synchronous in this allohexaploid than in its diploid parent species Secale cereale (Rs:Ds = 0.21) even though the DNA C value is 2.4-fold greater in the former than in the latter. Thus it was suggested that, like other aspects of DNA synthesis, the synchrony of activation of replicons, and hence the pattern of DNA synthesis, in rye chromosomes of triticale is controlled to be much more like that of wheat chromosomes. A similar phenomenon was noted previously (Bennett, Chapman & Riley 1971), since the duration of meiosis in octoploid triticale (21 h) was considered to be much more like that of hexaploid T. aestivum (24 h) than of diploid S. cereale (51.2 h). Such behaviour at S-phase and at meiosis is interesting because it suggests the existence in established wheat allopolyploids of an important genetic mechanism which normally acts to synchronise the developmental behaviour of their different ancestral diploid genomes. These mechanisms may also exercise a similar control over the behaviour of other genomes with different DNA C values or patterns of DNA synthesis (e.g. S. cereale), which have not previously been exposed to such control in an allopolyploid (NB there are no naturally occurring polyploid species in the genus Secale). Possession of such a mechanism may be essential to ensure nuclear stability in allopolyploids whose diploid donor genomes normally express different rates of development in the parent species, and may therefore characterize successful allopolyploids. Clearly, more allopolyploids must be analysed to test this hypothesis, and it will be important to compare (1) long-established successful allopolyploids with others newly derived from their parent diploids; and, (2) newly derived autopolyploids with their parent diploids. Such comparisons may help to distinguish between the effects of increased DNA amount and increased gene dosage due to polyploidy on the rate and pattern of activation of replicon families. In particular, the way in which replicons function in allo- as opposed to auto-polyploids may increase our understanding of DNA synthesis in relation to DNA C values, as well as the role of polyploidy in evolution.

ACKNOWLEDGEMENTS

We are grateful to the AFRC for supporting this work through grant AG 72/48.

REFERENCES

Bennett, M.D. (1971). The duration of meiosis. Proceedings of the Royal Society B, 178, 277-99.

Bennett, M.D. (1972). Nuclear DNA content and minimum generation time in herbaceous plants. Proceedings of the Royal Society of London B, 181, 109-35.

Bennett, M.D. (1976). DNA amount, latitude and crop plant distribution. Environmental and Experimental Botany, 16, 93-108.

Bennett, M.D. (1977a). The time and duration of meiosis. Philosophical Transactions of the Royal Society of London B, 277, 201-26.

Bennett, M.D. (1977b). Heterochromatin, aberrant endosperm nuclei and grain shrivelling in wheat-rye genotypes. Heredity, 39, 411-9.

Bennett, M.D. (1980). Theoretical and applied DNA studies and Triticale breeding. Hodowla Roslin Aklimatyzacja I Nasiennictwo, 24, 289-98.

Bennett, M.D., Chapman, V. & Riley, R. (1971). The duration of meiosis in pollen mother cells of wheat, rye and Triticale. Proceedings of the Royal Society of London B, 178, 259-75.

Bennett, M.D. & Finch, R.A. (1972). The mitotic cycle time of root meristem cells of Hordeum vulgare. Caryologia, 25, 439-44.

Bennett, M.D., Rao, M.K., Smith, J.B. & Bayliss, M.W. (1973). Cell development in the anther, the ovule, and the young seed of Triticum aestivum L. var. Chinese Spring. Philosophical Transactions of the Royal Society of London B, 266, 39-81.

Bennett, M.D. & Smith, J.B. (1972). The effects of polyploidy on meiotic duration and pollen development in cereal anthers. Proceedings of the Royal Society of London B, 181, 81-107.

Bennett, M.D. & Smith, J.B. (1976). Nuclear DNA amounts in angiosperms. Philosophical Transactions of the Royal Society of London B, 274, 227-74.

Blumenthal, A.B., Kriegstein, H.J. & Hogness, D.S. (1974). The units of DNA replication in Drosophila melanogaster chromosomes. Cold Spring Harbor Symposia in Quantitative Biology, 38, 205-23.

Bryant, J.A. (1982). DNA replication and the cell cycle. In Encyclopaedia of Plant Physiology 14B, ed. B. Parthier & D. Boulter pp.75-105. Berlin: Springer-Verlag.

Callan, H.G. (1972). Replication of DNA in the chromosomes of eukaryotes. Proceedings of the Royal Society of London B, 181, 19-41.

Cavalier-Smith, T. (1978). Nuclear volume control by nucleoskeletal DNA, selection for cell volume and cell growth rate, and the solution of the DNA C value paradox. Journal of Cell Science, 34, 247-78.

Davies, P.O.L. & Rees, H. (1975). Mitotic cycles in Triticum species. Heredity, 35, 337-9.

De Pamphilis, M.L. & Wassarman, P.M. (1980). Replication of eukaryotic chromosomes: A close-up of the replication fork. Annual Review of Biochemistry, 49, 627-66.

Evans, G.M. & Rees, H. (1971). Mitotic cycles in dicotyledons and monocotyledons. Nature, 233, 350-1.

Evans, G.M., Rees, H., Snell, C.L. & Sun, S. (1972). The relationship between nuclear DNA amount and the duration of the mitotic cycle. In Chromosomes Today eds. C.D. Darlington & K.R. Lewis, vol. 4 pp. 24-31. New York: Hafner.

Flavell, R. (1980). The molecular characterisation and organisation of plant chromosomal DNA sequences. Annual Review of Plant Physiology, 31, 569-96.

Francis, D. & Bennett, M.D. (1982). Replicon size and mean rate of DNA synthesis in rye (Secale cereale L. cv. Petkus Spring). Chromosoma, 86, 115-22.

Holm-Hansen, O. (1969). Algae: amounts of DNA and organic carbon in single cells. Science, 163, 87-8.

Huberman, J.A. & Riggs, A.D. (1966). Autoradiography of chromosomal DNA fibres from Chinese Hamster Cells. Proceedings of the National Academy of Science, 55, 599-606.

Huberman, J.A. & Tsai, A. (1973). Direction of DNA replication in mammalian cells. Journal of Molecular Biology, 75, 5-12.

Jones, R.N. & Rees, H. (1968). Nuclear variation in Allium. Heredity, 23, 591-605.

Kaltsikes, P.J. (1971). The mitotic cycle in an amphidiploid (Triticale) and its parental species. Canadian Journal of Genetics and Cytology, 13, 656-62.

Kornberg, A. (1980). DNA replication. San Francisco: Freeman.

Kowyama, Y. (1983). Cell cycle dependency of radio-sensitivity and mutagenesis in fertilized egg cells of rice, Oryza sativa L. 1. Autoradiographic determination of the first DNA synthetic phase; Theoretical and Applied Genetics, 65, 303-8.

Lewin, B. (1974). Gene expression 2: Eukaryotic chromosomes. New York: Wiley.

Lewin, B. (1980). Gene expression 2: Eukaryotic chromosomes, second edition. New York: Wiley.

Maher, E.P. & Fox, D.P. (1973). Multiplicity of ribosomal RNA genes in Vicia species with different nuclear DNA contents. Nature, New Biology, 245, 170-2.

Nagl, W. (1982). Nuclear chromatin. In Encyclopaedia of Plant Physiology, 14B, eds. B. Parthier & D. Boulter, pp. 1-45. Berlin: Springer-Verlag.

Sokurova, E.N. (1973). The content of nucleic acids in cells of yeasts belonging to various taxonomic groups. Mikrobiologiya, 42, 1020-4. (in Russian).

Sparrow, A.H., Price, H.J. & Underbrink, A.G. (1972). A survey of DNA content per cell and per chromosome of prokaryotic and eukaryotic organisms: Some evolutionary consideration. In Evolution of genetic systems ed. H.H. Smith pp. 451-93. New York: Gordon & Breach.

Thomas, C.A. (1971). The genetic organisation of chromosomes. Annual Review of Genetics 5, 237-56.

Van't Hof, J. (1965). Relationship between mitotic cycle duration, S period duration and the average rate of DNA synthesis in the root meristem cells of several plants. Experimental Cell Research, 39, 48-54.

Van't Hof, J. (1975). DNA fibre replication in chromosomes of a higher plant (Pisum sativum). Experimental Cell Research, 93, 95-104.

Van't Hof, J. (1976). DNA fibre replication of chromosomes of pea root cells terminating S. Experimental Cell Research, 99, 47-56.

Van't Hof, J. & Bjerknes, C.A. (1981). Similar replicon properties of higher plant cells with different S period and genome properties. Experimental Cell Research, 136, 461-5.

Van't Hof, J., Bjerknes, C.A. & Clinton, J.H. (1978). Replicon properties of chromosomal DNA fibres and the duration of DNA synthesis of sunflower root tip meristem cells at different temperatures. Chromosoma, 66, 161-71.

Van't Hof, J., Bjerknes, C.A. & Clinton, J.H. (1979). Replication of chromosomal DNA fibres of root meristem cells of higher plants. In Molecular Biology of Plants eds. H. Smith & D. Grierson, pp. 73-91. UK: Academic Press.

Van't Hof, J., Kuniyuki, A. & Bjerknes, C.A. (1978). The size and number of replicon families of chromosomal DNA of Arabidopsis thaliana. Chromosoma, 68, 269-85.

Van't Hof, J. & Sparrow, A.H. (1963). A relationship between DNA content, nuclear volume, and minimum generation time. Proceedings of the National Academy of Sciences, 49, 897-902.

Verma, R.S. (1980). The duration of G1, S, G2, and mitosis at four different temperatures in Zea Mays L. as measured with ^{3}H-thymidine. Cytologia, 45, 327-33.

CHROMATIN STRUCTURE, GENE EXPRESSION AND THE CELL CYCLE

P. Kelly and A.J. Trewavas

INTRODUCTION

Growth and development in plants and animals is governed by an interaction between a genetic programme and the effects of the environment. The mature individual is composed of a large number of different parts or organs, each of which has a specific role to play. The structure and function of each organ is in turn a reflection of the type or types of cells it contains, and different cell types are generally distinguishable both by morphology and metabolism. What is it, then, that makes for example, a mesophyll cell different from a phloem cell or dividing cells different from non-dividing cells? As proteins are involved in bringing about almost all metabolic and morphological changes in cells, then they would appear to be likely candidates; indeed it is now believed that most of the differences between cells are due to differences in the relative concentrations of the proteins they contain (Paul, 1982). As each of these proteins is coded for by a specific piece of DNA, or gene, it is relevant to ask how the expression of individual genes is controlled particularly, in the context of this volume, in relation to the cell cycle.

Most of the early work on gene expression involved prokaryotes, in particular the bacterium *E. coli* where control was found to be primarily at the level of transcription and brought about by specific effector molecules. However, in eukaryotes, the situation is much more complex, presumably as a consequence of the more complex organisation of the eukaryotic cell. For example, a eukaryotic cell may contain many thousand times more DNA than an *E. coli* cell, and unlike naked prokaryotic DNA, eukaryotic DNA is complexed extensively with protein and organised into chromosomes. Moreover, the presence of a nuclear membrane means that transcription and translation are separated in space and time in eukaryotes, whereas in prokaryotes they are closely coupled. Therefore gene expression in eukaryotes may involve, in addition to differential gene expression, mechanisms such as gene loss or amplification, DNA transposition or modification, RNA processing, transport, translation, or turnover and protein modification or turnover (Paul, 1982).

Thus the central dogma of molecular biology, "DNA transcribed to RNA, RNA translated to protein", must be expanded to include the above. Such a scheme is shown in Figure 1, where each step (arrow) represents a possible control point. Clearly, a review of this size cannot fully cover such a wide research area, and so we will confine ourselves to a discussion of control of transcription in eukaryotes, i.e. the first two steps of Figure 1. For a more detailed account of all aspects of gene expression, readers are referred to Lewin (1980). Firstly, we will consider, in more detail, RNA synthesis and the organisation of DNA in the nucleus.

SYNTHESIS OF RNA

In transcription, a molecule of RNA is synthesised on a DNA template. Thus the primary transcript has a nucleotide sequence complementary to a strand of the DNA duplex. Nuclear primary transcripts are, however, extremely short-lived, and are rapidly processed by selective removal and addition of nucleotide sequences (Breathnach & Chambon, 1981).

Transcription is carried out enzymatically by three types of RNA polymerase which vary in their template specificity. RNA polymerase I is located in the nucleolus and transcribes the genes coding for ribosomal RNA. Other non-protein-coding RNAs (eg. tRNA) are synthesised by polymerase III while primary transcripts which give rise to protein-coding mRNAs are synthesised using RNA polymerase II. Most of the following discussion concerns genes which are transcribed by RNA polymerase II.

In addition to being regulated, initiation of RNA synthesis shows specificity in that RNA polymerases only attach at selective sites on

Fig. 1. Flow diagram for gene expression in eukaryotes, with possible control points (boxed numbers). The curved arrows represent turnover.

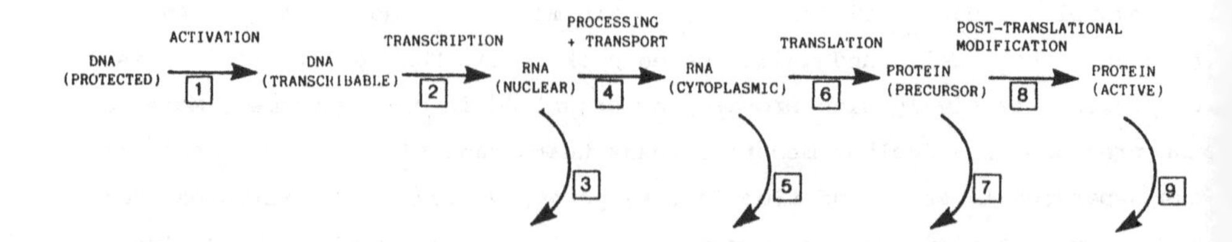

the DNA. Such sites, or promoters are believed to occur some 25-100 base pairs before the start of the DNA coding sequence (i.e. "upstream" of the transcription unit) (Lewis & Burgess, 1982). Nucleic acid sequencing of these regions has revealed at least three fairly well conserved "consensus" sequences, the Hogness box (Grosveld, De Boer, Shewmaker & Flavell, 1982) the Chambon box (Benoist, O'Hare, Breathnach & Chambon, 1980) and the "minus 100" region (Dierks _et al._, 1983). Studies using DNA base deletions and substitutions suggest that the Hogness (TATA) box, together with its flanking sequences, is the dominant element in determining the starting site of transcription, whereas the other two may be present to enhance or modulate initiation (Dierks _et al._, 1983). The approximate relative positions of these elements are shown in Figure 2. As highly purified RNA polymerase II cannot specifically recognise promoters, whereas crude preparations correctly initiate transcription, then co-factors or _transcription factors_ must be needed for accurate initiation. It was thought that such factors were general in action for all promoters and therefore had no role in differential gene expression, but recently Dynan and Tjian (1983) have isolated two transcription factors from HeLa cells, one of which is general in action, and the other which will selectively stimulate promotion of certain viral genes but not others.

The above example tentatively suggests that some differential control may be exerted by transcription factors in recognising promoters for specific genes, but whether this has relevance to cellular rather than virus genes remains to be seen. This would be an example of control at point

Fig. 2. Some consensus sequences in chromosomal DNA upstream of genes, thought to be important for accurate and efficient transcription. The numbers show the approximate distance in base pairs (bp) of the various elements from the start of the transcription unit.

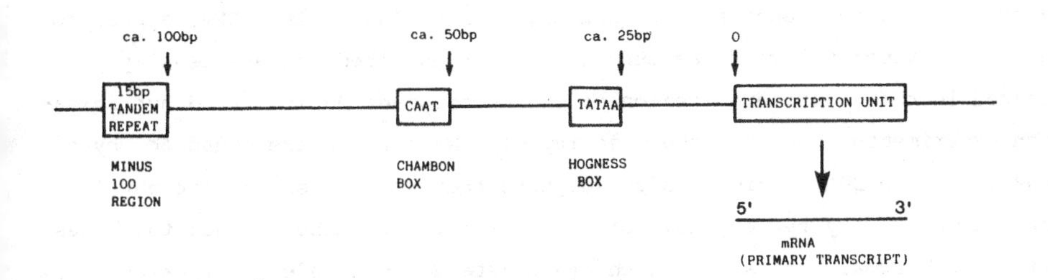

2 in Figure 1. The remainder of this article will be concerned with control at point 1, i.e. transcriptional activation of cellular DNA. It is therefore necessary to first of all consider the structural organisation of cellular DNA.

STRUCTURAL FEATURES OF CHROMATIN

The genetic material of eukaryote cells is concentrated largely, although not exclusively, in the nucleus. Cytoplasmic organelles such as chloroplasts and mitochondria also contain protein-coding DNA, but these are outside the scope of this article. Only about one half of the mass of eukaryote chromosomes is DNA, the rest being proteins. This nucleoprotein material is known as chromatin.

DNA

In prokaryotes, the genome size is compatible with the number of proteins they produce. Eukaryote cells, on the other hand, with about a thousand times (or more) as much DNA per genome are unlikely to have one thousand times as many genes, even allowing a liberal estimate of protein synthesising requirements. Another striking feature of eukaryote genome size is its wide variation, which is essentially unrelated to phenotypic complexity. Even closely related organisms can vary quite widely in haploid DNA content; for example, amphibians can have from 1 pg to 100 pg (Lewin, 1980), and the higher plant family Ranunculaceae shows an eighty-fold variation between species (Bennett & Smith, 1976)(see also Francis, Kidd & Bennett, this volume). How, then, does eukaryote DNA differ in organisation from that of prokaryotes, and how does this affect regulation of gene expression?

The DNA of any particular eukaryote can usually be categorised into three groups: non-repetitive, moderately repetitive and highly repetitive, by differing kinetics of reassociation following thermal denaturation. Highly repetitive sequences, such as "satellite" DNA, appear to have some structural role, as they are not transcribed and are usually located in the centromeric region of the chromosome. Nucleic acid hybridisation experiments show that the majority of mRNA species are coded for by the non-repetitive DNA portion, which suggests that most genes are present in the genome as very few copies. Notable exceptions to this include the genes for the ribosomal RNA family which are clustered and tandemly repeated many times.

Most "single-copy" genes are separated by moderately repetitive sequences of about 300 base pairs long, which may be involved in binding regulatory proteins and controlling the expression of adjoining structural genes (Davidson & Britten, 1979).

Histones

Chromatin proteins can be divided operatively into acid-soluble proteins (histones) and acid-insoluble proteins (nuclear acidic proteins). The histones, which constitute by far the largest group by mass, have an exceptionally high content of the basic amino acids lysine and arginine, and can be classified into five groups (H1, H2A, H2B, H3 and H4) according to their relative amounts of these amino acids. Various lines of evidence, including electron microscopy and nuclease digestion, have shown that histones interact with the DNA to form 200 base pair repeating units known as nucleosomes. Each unit contains a core of two molecules each of histones H2A, H2B, H3 and H4. Most of the DNA is wound around the core, and the remainder, or linker, joins adjacent nucleosomes and gives the chromatin fibre flexibility. Histone H1, on the other hand, is not always present in nucleosomes, and when present has a stoichiometry of only one per nucleosome. It also appears to have a peripheral location rather than being part of the core. Reconstitution studies suggest that H1 may act as a bridge between nucleosomes and help to give the chromatin fibre a compact helical structure of 100Å diameter (Fig. 3).

Addition of low concentrations of Mg^{2+} (1mM) to this 100Å fibre *in vitro* results in the formation of a solenoidal structure with six to seven nucleosomes per turn and a width of around 250-300 Å (Finch & Klug, 1976) which is the observed chromatin fibre diameter (Fig. 3).

Non-histone proteins

Nuclear "acidic" proteins are in fact only weakly acidic, and generally account for less than 15% of the chromatin mass. Individual proteins are variable in abundance and the major species are generally structural, such as tubulin, actin, and components of the nuclear matrix. Others are thought to serve a regulatory role and will be discussed in the next section while yet others may be enzymes. Several non-histone proteins are capable of being phosphorylated by nuclear-based protein kinase.

ACTIVE AND INACTIVE CHROMATIN

Generally, individual chromosomes can only be observed in the eukaryotic nucleus during the middle to late stages of cell division (mitosis). Between division (interphase) the chromatin exists in a more diffuse state, with the degree of condensation being variable.

The more condensed forms, or heterochromatin, stain densely with certain dyes, replicate late in S-phase, and are inactive in RNA synthesis. Euchromatin, on the other hand, is diffuse, stains lightly, and is active in RNA synthesis. The relative amounts of each vary with cell types, and it is possible that a conversion from dense to diffuse chromatin occurs at times of high RNA synthesis (see Nagl & Scherthan, this volume). The transition is accompanied by a change in chromatin fibre diameter, and an unravelling of the whole chromatin structure (Jordan, Timmis & Trewavas,

Fig. 3. A diagram showing the various levels of chromatin structure and their interrelationships. The numbers on the DNA of the extended nucleosomes indicate the lengths of core (140 base pair) and linker (approximately 60 base pair) DNA. The numbers on the 110 and 250 Å fibres refer to adjacent nucleosomes. For clarity the DNA has been omitted from the 250 Å fibre. From Jordan et al. 1980, by kind permission of Academic Press, Inc.

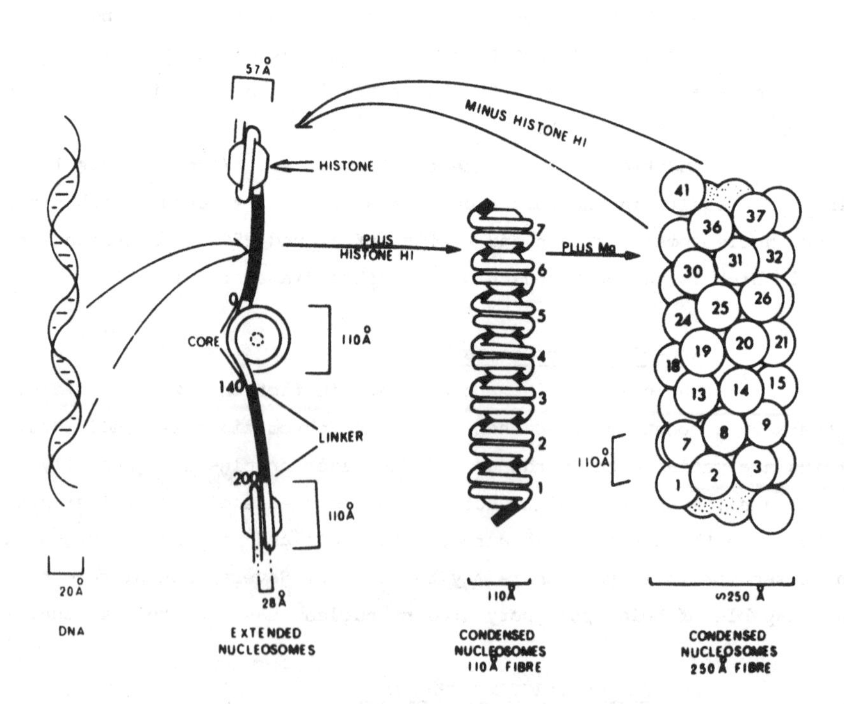

1980). This may be analogous to the effect of removal of histone H1 shown in Fig. 3, and certainly isolated diffuse chromatin is depleted in H1, but this may be an artefact of preparation (Lewin, 1980).

Some of the most direct early evidence linking diffuse chromatin with areas of transcription came from studies of Dipteran polytene chromosomes (see Lewin, 1980). These large banded structures are found in the salivary gland of flies such as *Drosophila*, and contain more than one thousand aligned DNA molecules. During development of a larva into a pupa, certain bands can be seen to become diffuse and form *puffs*. These puffs have been shown by autoradiography to be active in RNA synthesis. Specific bands become diffuse and then contract in a precise temporal sequence. The whole process appears to require the continuous presence or even be under the specific control of the insect hormone *ecdysone*.

What species of molecules, then, are involved in distinguishing regions of chromatin which are transcribed from those which are not? Attempts have been made to fractionate chromatin by differing physical properties, but these have so far not lead to the identification of any changes in chromatin structure with transcriptional state. However, more sensitive fractionation methods may be developed in the future to resolve any differences (Lewin, 1980).

One important consequence of the transition to diffuse chromatin is that it becomes more accessible to small molecules. This has been clearly demonstrated by the use of nucleases; treatment of chromatin with such enzymes leads to the selective release of diffuse chromatin (Marushige & Bonner, 1971), thus indicating that the DNA of diffuse chromatin is more accessible to enzymic hydrolysis. Differences in sensitivity have also been shown in individual genes involved in transcription. For example, the globin gene is actively transcribed in the immature chick erythrocyte nucleus but not in the fibroblast nucleus. Using nucleases, Weintraub and Groudine (1976) demonstrated that the globin gene in erythrocyte chromatin was very susceptible to digestion while that in the fibroblast nucleus was not.

Further work from Weintraub's group showed a differential DNase sensitivity of embryonic and adult β-globin genes during chick red blood cell development, (Stalder *et al.*, 1980a). Also, a lesser DNase sensitivity extends to regions up to 8000 bases, both upstream and downstream of the highly sensitive coding regions (Stalder *et al.*, 1980b). In addition, sites which are hypersensitive to the nuclease DNase I have been found at or near the 5' end of active genes. This sensitivity has not been found in tissues where a particular gene is inactive (reviewed in Elgin, 1981). Indeed, it seems likely that a DNase-I-sensitive site is a necessary but not sufficient condition for transcription by RNA polymerase II *in vivo*. Such specific sites are not found in naked DNA, and must therefore be determined by the overall structure of the chromatin fibre. Perhaps an unusual DNA structure or a local absence of nucleosomes could be generated. In any case, the site may well have a regulatory role in transcription.

CHROMATIN PROTEINS AS CONTROLLING ELEMENTS

Some early theories on transcriptional activation suggested that histones coating the DNA sterically blocked RNA synthesis, and that active genes may be depleted in histones. However, it soon became clear that the histones lack the specificity to regulate gene expression. Moreover, nuclease-released diffuse chromatin, and also individual transcribed genes, have been shown to be in the form of nucleosomes (Jordan *et al.*, 1980). Thus it seems likely that histones fulfil a more general role in the organisation of chromosomes rather than being involved in gene regulation. However, the role of H1 binding in control of transcription is still an open question.

Nuclear acidic proteins, on the other hand, are believed to be much more variable, quantitively and qualitively, and thus better candidates for controlling elements. Indeed, it is often claimed that the proportion of non-histone protein can be two to four-fold higher in diffuse than in dense chromatin, and that this difference may determine whether transcription takes place, but this conclusion is open to criticism (Lewin, 1980).

Firstly, the fraction isolated as "non-histone protein" is often derived from whole nuclei rather than chromatin and therefore may include many proteins not directly associated with chromatin. Secondly, comparison of non-histone proteins from different sources by gel electrophoresis generally shows strong similarities between related chromatins; in dense and diffuse chromatin most of the gel bands are the same, and minor changes, detected by low-resolution scanning techniques, may be artefactual. In other words, gross differences in non-histone protein composition between dense and diffuse chromatin appear unlikely, and a higher resolution analysis is needed to clearly identify minor differences.

Some apparently better evidence for a role of non-histone proteins comes from chromatin reconstitution experiments (reviewed by Jordan *et al.*, 1980). In these, the chromatin from two related cell types, only one of which expresses a particular gene, is divided into its "three components" i.e. DNA, histones and non-histone proteins. Heterologous chromatins are then made by combining these fractions, the reconstituted chromatin is transcribed in vitro, and an assay made for the specific transcript. Using this method, Gilmour and Paul (1973) showed that genetic restriction of the globin gene was a property of the tissue source of the non-histone protein rather than that of the DNA or histone. Similar results have been reported for a few other genes (see Jordan et al., 1980).

However, the interpretation of the results of these experiments has been seriously questioned by Lewin (1980) who has pointed out that the pattern of transcription in vitro may not resemble that of the tissue from which the chromatin was obtained, and that the transcription assay is probably not sensitive enough to distinguish faithful from random transcription. The main weakness in the in vitro transcription experiments is that they use added bacterial RNA polymerase, or occasionally purified mammalian or plant RNA polymerase II (i.e. without transcription factors, see above), neither of which are noted for their ability to recognise promotors and thereby initiate specific, faithful transcription. For example, in cases where the strand symmetry of transcription has been followed, both strands rather than one specific strand of DNA, are copied. The criticism of the assay is more technical; briefly, in vitro transcribed RNA species are hybridised to an excess of a cDNA probe (a DNA "copy" of a specific mRNA) and the amount of hybrid molecules measured. Such cDNA probes, synthesised by reverse transcription of mRNA from its 3' end, are often short. Therefore the assay in fact detects transcription in vitro of sequences at the 3' end of the gene. If initiation of transcription was specific then these transcripts must be large; however, most evidence shows that in vitro transcripts are short, therefore the assay must be measuring non-specific products from the far end of the transcription unit, and not faithful transcripts.

More definite evidence for the involvement of specific non-histone proteins in expressed chromatin regions has been provided by partial nuclease digestion, which preferentially releases certain non-histone proteins. Two in particular, HMG14 and HMG17, have been implicated in the changed DNase sensitivity in the region around transcribed globin genes in chick erythrocytes (Weisbrod, Groudine & Weintraub, 1980). Both proteins are associated with the nucleosome core (Igo-Kemenes, Horz & Zachau, 1982) and will not bind even to an excess of "inactive" nucleosomes. In other words, DNase I sensitivity is conferred on, for example, the globin genes but not the ovalbumin gene of red blood cells, the latter gene being inactive in these cells. Therefore, active nucleosomes have at least one unique but unknown property which directs HMG binding. The general consensus is that HMG 14 and 17 occupy two binding sites at each end of the nucleosome core and interact with approximately 150-160 base pairs of DNA, possibly replacing histone H1 and giving a more open chromatin fibre (Weisbrod, 1982).

MODIFICATION OF CHROMATIN PROTEINS

Many of the proteins of chromatin undergo post-translational modifications. Of these, the most studied are phosphorylation and acetylation. Phosphorylation of certain histone and non-histone proteins is associated with chromatin condensation during the cell cycle, and will be discussed below. Although active chromatin is often rich in acetylated histones, this acetylation does not appear to be necessary to enhance either DNase I sensitivity or the rate of transcription (Weisbrod, 1982). Therefore, the specific function of high histone acetylation in active genes remains unknown.

DNA METHYLATION

Although methylated bases are fairly common in plant DNA, the only well characterised methylated base in mammalian DNA is 5-methylcytosine (^{m}C), which is predominantly found in the sequence 5'CpG3'. The modification is stable and clonally inherited, but tissue specific methylation patterns are also created during development. The latter have been inversely correlated with gene activation, or in other words, specific sites appear to be under-methlated in active genes (review by Razin & Riggs, 1980). The position of methylated bases is important if gene expression is to be inhibited; for example, Busslinger, Hurst & Flavell (1983), using an *in vitro* methylation technique, found that relative to the start of the transcription unit, ^{m}C residues from -760 base pairs (i.e. upstream in the 5' flanking region) to +100 base pairs in the γ-globin gene prevent transcription.

The conversion of C to ^{m}C introduces a methyl group into the major groove of the DNA and changes in this groove are known to affect the binding of certain proteins to DNA. It has therefore been postulated that methylation affects gene activity by modifying protein-DNA interactions (Razin & Riggs, 1980), but this has not been proved.

However, the correlation is not universal, i.e. not all developmentally de-methylated genes are active, and also some heavily methylated genes are expressed. For example, the estrogen-induced vitallo-genin genes in *Xenopus* hepatocytes are expressed when totally methylated (Gerber-Huber *et al.*, 1983). Therefore, changes in the methylation pattern may not be a completely general prerequisite for gene activation.

CHROMATIN IN THE CELL CYCLE

The life cycle of a dividing cell can be categorised into two parts by observation of the morphology of its chromatin; these are mitosis and interphase. In mitosis, during which the cell divides, chromosomes are visible, whereas in interphase the chromatin is observed as a diffuse network. Between the end of one mitosis and the beginning of the next, the diploid amount of genetic material must double. As DNA synthesis only takes place during a restricted period, interphase can be normally divided operationally into three parts, i.e. two "gaps" (G1 and G2) before and after the period of DNA synthesis itself, S-phase. Cells that cease to divide and remain in G1 are often referred to as GO cells. However GO may simply be a special and reversible example of the G1 state.

The appearance of visible chromosomes at the onset of mitosis is due to condensation of the chromatin, and requires a further two orders of magnitude of compaction above that of the 250 Å "solenoid" to produce the characteristic paired, and positioned chromosome morphology of metaphase (mid-mitosis). Removal of histones from HeLa metaphase chromosomes produces long loops of DNA, or "domains" of about 40 kilobases, which extend out from a core structure or scaffold. Removal of the DNA leaves a non-histone protein structure which retains the morphology of the metaphase chromosomes (Paulson and Laemmli, 1977). Recently a protein, perichromin, has been identified which is present on the periphery of metaphase chromosomes, and is conserved in the animal kingdom from mammals through to *Drosophila* (McKeon, Tuffanelli, Kobayashi & Kirshner, 1984). This protein, however, is also present during interphase, at the surface of the nuclear envelope. This location suggests a pathway and possibly a role for perichromin, in organising the transition from interphase to metaphase chromosome morphology. It also suggests that the "domain" structure may have some relevance to interphase chromosomes, which are actively transcribed, as well as to metaphase chromosomes which are not.

Another significant feature found during chromatin condensation prior to mitosis is a massive phosphorylation of histone H1 (reviewed by Bradbury & Matthews, 1982). In *Physarum*, for example, the amount of phosphorylated H1 increases three-to-six-fold during middle to late G2 and decreases sharply late in mitosis (see also Waterborg & Shall, this volume).

The level of specific H1 kinase also rises in late G2. It is not known, however, how this phosphorylation can help in chromosome condensation, particularly as the simple electrostatic effect of phosphorylation is to reduce the strength of H1 binding to DNA. Similar observations have been made in higher plants (Stratton & Trewavas, 1981). Both HMG14 and HMG17 also undergo phosphorylation prior to mitosis, suggesting that they also may have a role in condensation and gives weight to the theory that they may substitute for histone H1 in regions of active chromatin (Weisbrod, 1982).

The synthesis of histones also seems to be cell-cycle-dependent. The whole chomosome, not just the DNA, is reproduced during S-phase and isotope incorporation experiments show *de novo* synthesis of both histone proteins and their mRNAs during this time (Lewin, 1980). Histone synthesis may continue even when DNA synthesis is blocked, suggesting only a weak coupling between the two processes. In some cells histone synthesis can occur to a limited extent at other times in the cell cycle (Jordan *et al.*, 1980; Lewin, 1980).

In general, however, there are apparently no other major cell-cycle-dependent qualitative changes in gene expression. Certain changes in non-histone protein components of the nucleus occur, and some of these may be due to *de novo* protein synthesis. Chafouleas *et al.*, (1984) have recently reported that changes in calmodulin and its mRNA accompany the re-entry of quiescent (G0) cultured cells into the cell cycle. Their results suggest that the calcium-calmodulin system may have a regulatory role at several points in the cell cycle (see Trewavas, this volume).

CONCLUSIONS AND OUTLOOK

In this short review, we have tried to outline the main approaches which have been made towards understanding transcriptional control, particularly in relation to gene expression during the cell cycle. In general, however, the question of how gene expression is controlled is still unanswered. Certainly the distinctive structural features of active chromatin described above, such as decondensation of chromosomes, histone modifications, changes in non-histone protein composition, nuclease sensitivity, the existence of specific nuclease sensitive sites, binding of HMG14 and 17, and undermethylation, show a correlation with gene activity.

Some of these features are necessary but not sufficient for a gene to be active, but which of these has a causal role is unknown. In fact, there may be several different mechanisms involved in activating genes, to take account of differential amounts of expression. The total number of expressed genes in a single mammalian, avian or plant cell may be 5,000-10,000 (Lewin, 1980). Most of these are "housekeeping" genes whose products appear to be necessary for sustenance of all cell types. However, the cellular levels of the mRNAs for these proteins is much lower than those for specialised or "luxury" proteins such as legume storage proteins (Goldberg, 1983) in the appropriate cells (e.g. storage parenchyma of seed cotyledons).

Therefore, it is not simply a question of determining whether or not a gene is active, but also its degree of activity. Paul (1982) has calculated (using average estimates of elongation rate and mRNA half-life) that it only requires about two to three transcripts per gene per hour to maintain a steady state mRNA level of 10 to 20 molecules per cell. This would suggest either that the elongation rate is extremely low, or more likely that specific genes are only transcribed for a fraction of the time. The latter may indicate that initiation of transcription is the rate limiting factor, and it is tempting to speculate that chromatin features upstream from the 5'-end of the transcription unit may be involved in modulating the rate of initiation.

In any case, it is certainly obvious from the above considerations that to understand gene expression, we must analyse the structure and function of chromatin and not simply its DNA. Newer techniques such as improved fractionation of active chromatin using HMG-agarose affinity columns (Igo-Kemenes *et al.*, 1982), *in vitro* methylation techniques (Busslinger *et al.*, 1983), restriction enzymes which distinguish methylated bases (Bird and Southern, 1978) and commercially available *in vitro* transcription systems, should help this analysis greatly.

REFERENCES

Bennett, M.D. & Smith, J.B. (1976). Nuclear DNA amounts in angiosperms. *Philosophical Transactions of the Royal Society of London, Series B*, 274, 222-74.

Benoist, C., O'Hare, K., Breathnach, R. & Chambon, P. (1980). The ovalbumin gene-sequence of putative control regions. *Nucleic Acids Research*, 8, 127-42.

Bird, A.P. & Southern, E.M. (1978). Use of restriction enzymes to study eukaryotic DNA methylation. 1. The methylation pattern in ribosomal DNA of *Xenopus laevis*. *Journal of Molecular Biology*, 118, 27-47.

Bradbury, E.M. & Matthews, H.R. (1982). Chromatin structure, histone modifications, and the cell cycle. In: *Cell Growth*, ed. C. Nicolini, pp. 411-54. New York: Plenum.

Breathnach, R. & Chambon, P. (1981). Organisation and expression of eukaryotic split genes coding for proteins. Annual Review of Biochemistry, 50, 349-83.

Busslinger, M., Hurst, J. & Flavell, R.A. (1983). DNA methylation and the regulation of globin gene expression. Cell 34, 197-206.

Chafouleas, J.G., Lagace, L., Bolton, W.E., Boyd, A.E. & Means, A.R. (1984). Changes in calmodulin and its mRNA accompany re-entry of quiescent (GO) cells into the cell cycle. Cell, 36, 73-81.

Davidson, E.H. & Britten, R. (1979). Regulation of gene expression: possible role of repetitive sequences. Science, 204, 1052-9.

Dierks, P., van Ooyew, A., Cochran, M.D., Dobkin, C., Reiser, J. & Weissmann, C. (1983). Three regions upstream from the cap site are required for efficient and accurate transcription of the rabbit β-globin gene in mouse 3T6 cells. Cell, 32, 695-706.

Dynan, W.S. & Tjian, R. (1983). Isolation of transcription factors that discriminate between different promotors recognised by RNA polymerase II. Cell, 32, 669-80.

Elgin, S.C.R. (1981). DNase I - hypersensitive sites of chromatin. Cell, 27, 413-5.

Finch, J.T. & Klug, A. (1976). Solenoidal model for superstructure in chromatin. Proceedings of the National Academy of Science, USA., 73, 1897-1901.

Gerber-Huber, S., May, F.E.B., Westley, B.R., Felber, B.K., Hosbach, H.A., Andres, A.-C. & Ryffel, G.U. (1983). In contrast to other Xenopus genes, the estrogen-inducible vitellogenin genes are expressed when totally methylated. Cell, 33, 43-51.

Gilmour, R.S. & Paul, J. (1973). Tissue specific transcription of the globin gene in isolated chromatin. Proceedings of the National Academy of Science, USA., 70, 3440-2.

Goldberg, R.B. (1983). Organization and expression of soybean seed protein genes. In: Genetic Engineering: Applications to Agriculture, ed. L.D. Owens, pp. 137-50. Totowa, N.J./St. Albans, U.K.: Rowman & Allanheld/Granada.

Grosveld, G.C., De Boer, E., Shewmaker, C.K. & Flavell, R.A. (1982). DNA sequences necessary for transcription of the rabbit β-globin gene in vivo. Nature, 295, 120-6.

Igo-Kemenes, T., Horz, W. & Zachau, H.G. (1982). Chromatin. Annual Review of Biochemistry, 51, 89-121.

Jordan, E.G., Timmis, J.N. and Trewavas, A.J. (1980). The plant nucleus. In: The Biochemistry of Plants, Vol. 1, ed. N.E. Tolbert, pp. 489-588. New York and London: Academic Press.

Lewin, B. (1980). Gene Expression 2: Eukaryotic Chromosomes, second edition, New York: Wiley-Interscience.

Lewis, M.K. & Burgess, R.R. (1982). Eukaryotic RNA polymerases. In: The Enzymes, 3rd edition, Vol. 15B, ed. P.O. Boyer, pp. 109-53. New York: Academic Press.

McKeon, F.B., Tuffanelli, D.L., Kobayashi, S. & Kirshner, M.W. (1984). The redistribution of a conserved nuclear envelope protein during the cell cycle suggests a pathway for chromosome condensation. Cell, 36, 83-92.

Marushige, K. & Bonner, J. (1971). Fractionation of liver chromatin. Proceedings of the National Academy of Sciences, USA, 68, 2941-4.

Paul, J. (1982). Transcriptional control during development. Bioscience Reports, 2, 63-73.

Paulson, J.R. & Laemmli, U.K. (1977). The structure of histone-depleted metaphase chromosomes. Cell, 12, 817-28.

Razin, A. & Riggs, A.D. (1980). DNA methylation and gene function. Science, 210, 604-10.

Stalder, A., Groudine, M., Dogson, J.B., Engel, J.D. & Weintraub, H. (1980a) Hb switching in chickens. Cell, 19, 973-80.

Stalder, A., Larsen, A., Engel, J.D., Dolan, M., Groudine, M. & Weintraub, H. (1980b). Tissue-specific DNA cleavages in the globin chromatin domain introduced by DNase I. Cell, 20, 451-60.

Stratton, B.R. & Trewavas, A.J. (1981). Phosphorylation of histone H1 during the cell cycle of artichoke. Plant, Cell and Environment, 4, 419-26.

Weintraub, H. & Groudine, M. (1976). Chromosomal subunits in active genes have an altered conformation. Science, 193, 848-56.

Weisbrod, S. (1982). Active chromatin. Nature, 297, 289-95.

Weisbrod, S., Groudine, M. & Weintraub, H. (1980). Interaction of HMG14 and 17 with actively transcribed genes. Cell, 19, 289-301.

CHANGES IN CHROMATIN STRUCTURE DURING THE CELL CYCLE

W. Nagl and H. Scherthan

INTRODUCTION

Until recently, no distinction could be made between interphase nuclei at different stages of the cell cycle, differentiating nuclei and terminally differentiated nuclei. In 1963, however, Mazia suggested that chromosomes undergo a condensation - decondensation sequence extending throughout the entire mitotic cycle with two extreme conformational states - mitosis (condensed) and S-phase (decondensed). Since this time, much finer details have been revealed by various techniques, which now allow identification of the stages of the cell cycle by means of chromatin structure at different levels. This is particularly important for the differentiation of nuclei in G1, S, and G2 of the mitotic cycle, as the stages of mitosis itself can be distinguished from each other without any problem.

QUALITATIVE ANALYSIS OF CHROMATIN STRUCTURE

Identification of nuclei in various stages of interphase is possible by qualitative criteria as appears in bright field light microscopy, fluorescence microscopy, and electron microscopy. Following publications of Altmann (1966) and Müller (1966) on changes of nuclear structure during the cell cycle in liver epithelium cells of the mouse, I started structural analyses of nuclei in meristematic root tip cells of *Allium carinatum* and *A. flavum*. The different patterns identified could be assigned to certain cell cycle stages with the aid of cytophotometric DNA measurements, ^{3}H-thymidine autoradiography, and measurements of nuclear volume (Nagl, 1968, 1970a, b). The euchromatin of G1 nuclei appears in the form of granules (chromomeres) and fibrils (chromonemata) of an average diameter of 0.25 µm, that of G2 nuclei exhibits diameters between 0.4 and 0.5 µm. The heterochroma tin appears in the form of chromocenters, the size of which is increased, but the number of which is reduced, during interphase due to fusion of neighbouring ones. Similar changes in chromatin organization have been found in several other species; particularly clear alterations between G1 and G2 are visible in DNA-(chromatin-) rich species, such as Hyacinthus and Tradescantia. In unstained semi-thin sections, phase-contrast microscopy reveals a very fine chromatin structure in G1, but a rather rough one in G2. The transitions during the S-phase are sequential so that no simple description is possible. Fig. 1 summarizes the changes occurring between G1 and G2 in the nuclei belonging to the main structural types.

Few fluorescence microscopic analyses of cycling cells have been performed in plants, although flow cytometry can be used for cell cycle studies on the basis of the nuclear DNA content (Heller, 1973; Galbraith et al., 1983). Most data are from mammalian cells, using quantitative DNA analysis by flow cytophotometry (see the following section). However, staining with base-specific fluorochromes (e.g. DAPI) resolves various aspects of nuclear cycles in plants. For instance, heterochromatin amplification during the polyploidization cycle could be visualised by staining with AT-

Fig. 1. Diagrams to illustrate the characteristic changes in nuclear structure between G1 and G2, as seen in the light microscope. (a) chromocentric nuclei with diffuse euchromatin; (b) chromomeric nuclei; (c) chromonematic (reticulate) nuclei.

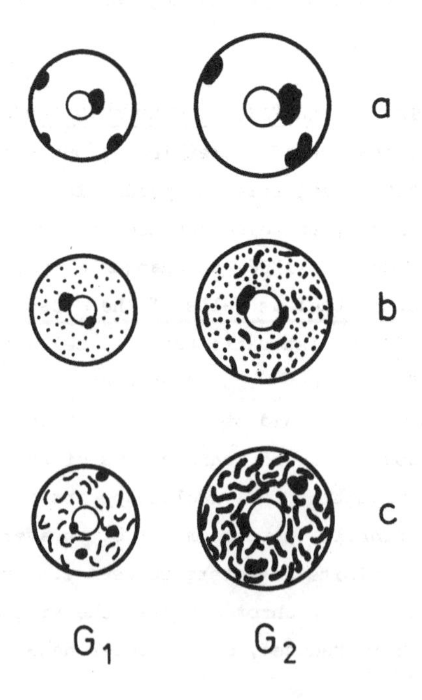

specific fluorochromes (Schweizer & Nagl, 1976). The results were consistent with data obtained by Feulgen cytophotometry (Nagl 1972, 1977) and by biochemical analyses of DNA (Nagl & Rücker, 1976; Nagl, Pohl & Radler, this volume).

Recently, chromatin organization in cycling plant cell nuclei was studied at the electron microscope level. The gross structure of nuclei in angiosperms depends on the basic nuclear DNA content per diploid genome (2*C* value), whereby the repetitive DNA fractions apparently play the major role in determining the degree of chromatin condensation (Nagl, 1982; Nagl *et al.*, 1983). The low-DNA species display a completely diffuse euchromatin, species with higher DNA values exhibit a chromomeric structure, and those with the highest DNA amounts a chromonematic organization of the euchromatin (see Fig. 1). Heterochromatin in the form of chromocenters may be present in addition, or may be missing. Interphasic changes of the nuclear ultrastructure are particularly clear in chromonematic nuclei (Figs. 2, 3). The increase in the amount and size of condensed chromatin regions is the result of DNA (chromatin) replication during the S period, and the fact that the nuclear volume increases by only about 60 per cent during interphase. The complete doubling of the nuclear volume is not reached before early prophase.

QUANTITATIVE ANALYSIS OF CHROMATIN STRUCTURE

By automated image analysis it has become possible to digitize and objectively characterize the changes in chromatin structure during the cell cycle (Sawicki, Rowinski & Abramczuk, 1974; Kendall *et al.*, 1977; Nicolini, Ajiro, Borun & Baserga, 1975; Nicolini, Kendall & Giaretti, 1977). The main methods employed are fluorescence staining, flow cytophotometry and scanning densitometry, but with other physical and chemical techniques such as analysis of thermal stability, light scattering, molecular ellipticity, circular dichroism and nuclease digestion. (For reviews see Nicolini, 1979, 1980).

Mitchell, Cohn & Van der Ploeg (1983) described quantitative changes in the degree of chromatin condensation during the cell cycle in differentiating vascular tissue of *Pisum sativum*, using scanning cytophotometry of Feulgen-stained nuclei by setting the instrument's detection threshold for "condensed chromatin" for absorbance values greater than 0.32. They found that the percentage of condensed chromatin decreased from about 45% to 12% during G1, increased rapidly at the end of S-phase, decreased

Fig. 2. Changes in nuclear ultrastructure during the interphase of maize (Zea mays) root tip cells. (a) very early G1, the telophasic arrangement of chromatin is still visible; (b) G1 phase, (c) G2 phase. (x 10,000).

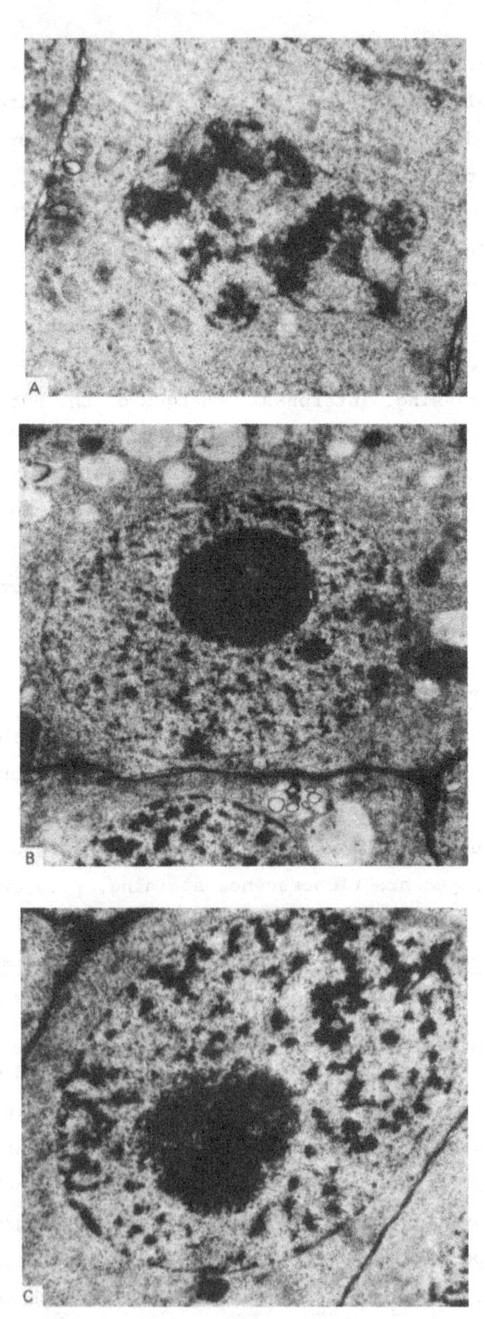

Fig. 3. Ultrastructural differences in chromatin organization between (a) G1 and (b) G2 in root tip nuclei of _Phaseolus coccineus_. Here, mainly the increase in the size of chromocenters is visible. (x 15,000).

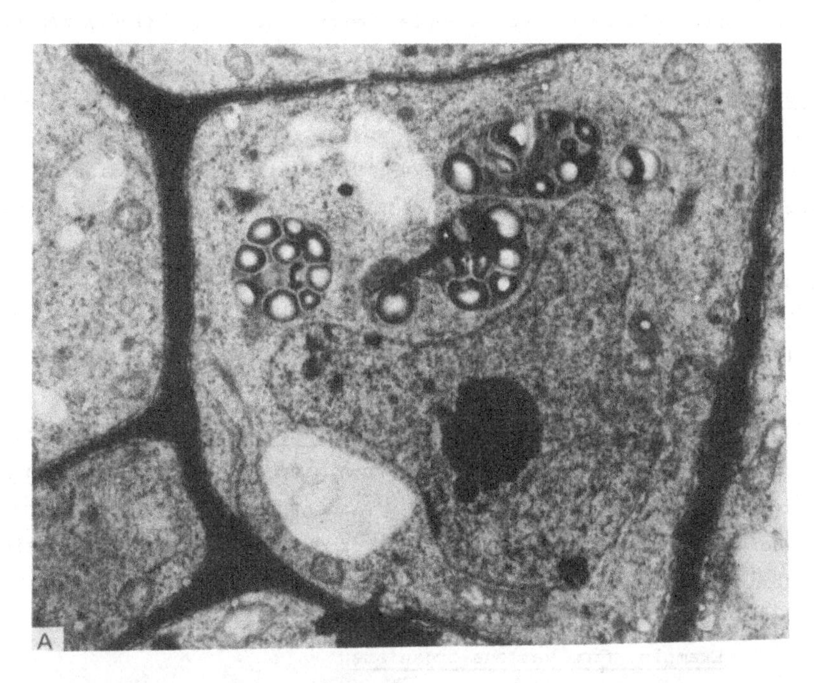

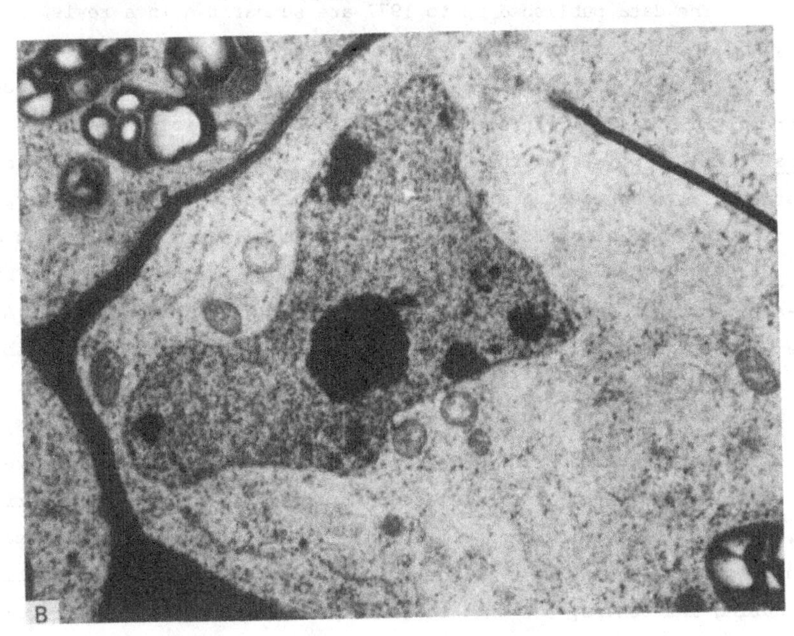

during G2 and the following G1, to increase again during the subsequent S-phase (Fig. 4).

Structural changes can also be digitized and quantified at the electron microscopic level. For example, Lafontaine & Lord (1969) reported an increase of the chromonema diameter between G1 and G2 in *Triticum aestivum* nuclei from about 0.15 to 0.3 µm. Similar changes were found in *Allium porrum* (Lafontaine & Lord, 1974) and *Allium fistulosum* (Klueva, Onischchenko, Poliakov & Chentsow, 1974). Fibre size, and the distribution of fibres within G1, S and G2 nuclei of *Allium cepa* were estimated by De la Torre, Sacristán-Gárate & Navarrete (1975) using stereological methods. These authors found evidence for a cyclic pattern of chromatin condensation during the cell cycle. Using automatic and semi-automatic image analysing systems, Nagl (1983) found an increase in the proportion of condensed chromatin from about 31% to about 38% in embryonic cell nuclei of *Lathrus odoratus*, and from about 20% to about 40% in root tip cells of *Hyacinthus orientalis* (Fig. 5). Fig. 6 displays the densitograms of electron micrographs of *Crepis cappill-aris* nuclei in G1, S and G2 phase. These results show that objective criteria exist whereby an automated identification of cell cycle stages is possible in structural terms.

Examples from various organisms

The data published up to 1977 are summarized in a review by Nagl (1977), and more recent data in a review by Nagl (1982).

Amongst plants, chromatin structural changes could be detected in extremely different organisms, such as algae and angiosperm species. In *Euglena gracilis*, Moyne, Bertaux & Puvion (1975) showed some changes in the ultrastructural organization of the condensed chromosomes during interphase. Studying successive stages of development of the antheridial filaments of *Chara vulgaris*, Olszewska & Marciniak (1977) found differences in the binding of ^{3}H-actinomycin D during the cell cycle and during differentiation. A similar investigation, using acridine orange as a probe for DNA availability was made on the first cell cycle in meristematic root cells of germinating *Pisum sativum* by Troyan *et al*. (1984).

Mammalian cells have been particularly well studied (for reviews see Nicolini, 1979, 1980; for an electron microscopic study see Setterfield *et al*., 1983). In certain synchronized cell lines it could be shown that the condensation pattern does not follow a simple decondensation-condensation cycle during interphase, but that rhythmic changes take place, particularly

Fig. 4. Changes in the percentage of condensed chromatin during the interphase of xylem cells in differentiating pea (*Pisum sativum*) internodes. The left plot represents G1 and G0 nuclei, the right plot S and G2 nuclei (in endo-cycles). Measurements were made on Feulgen-stained nuclei with a scanning cytophotometer interfaced to a PDP-11 computer. (Redrawn and modified from Mitchell *et al.*, (1983)).

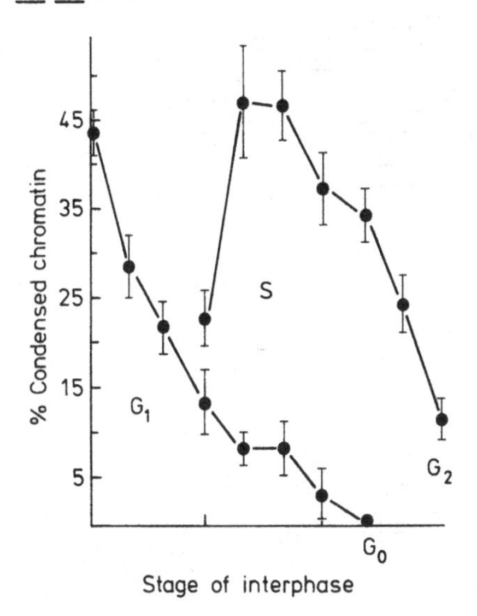

Fig. 5. Changes in the percentage of condensed chromatin during the cell cycle in *Hyacinthus orientalis* as measured in electron micrographs by the semi-automatic equipment MOP-Digiplan. (Redrawn from Nagl, (1983)).

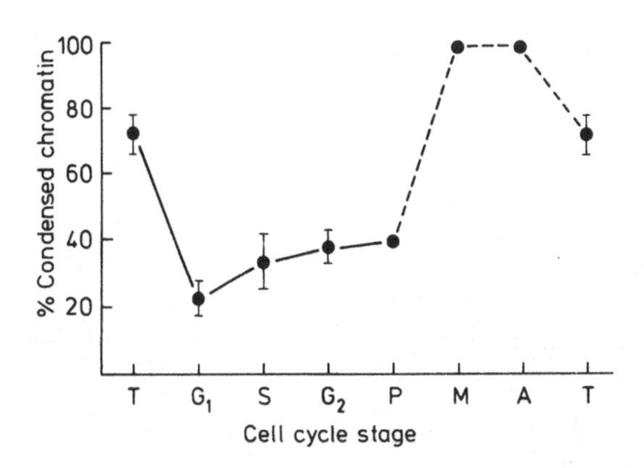

Fig. 6. Densitograms of electron micrographs of nuclei in G1, S and G2 phase in Crepis capillaris. Computer graphs taken with the Leitz texture analysing system, TAS.

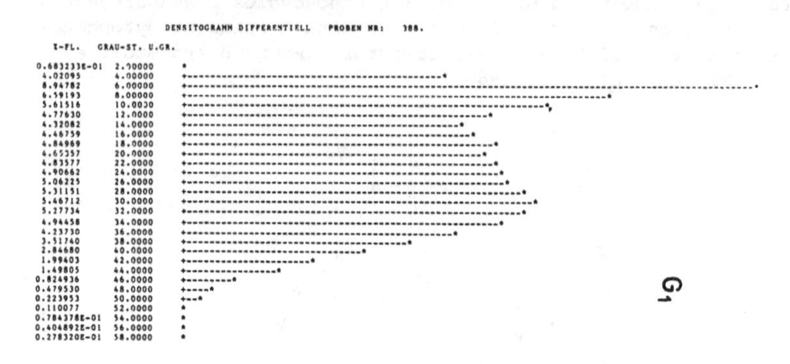

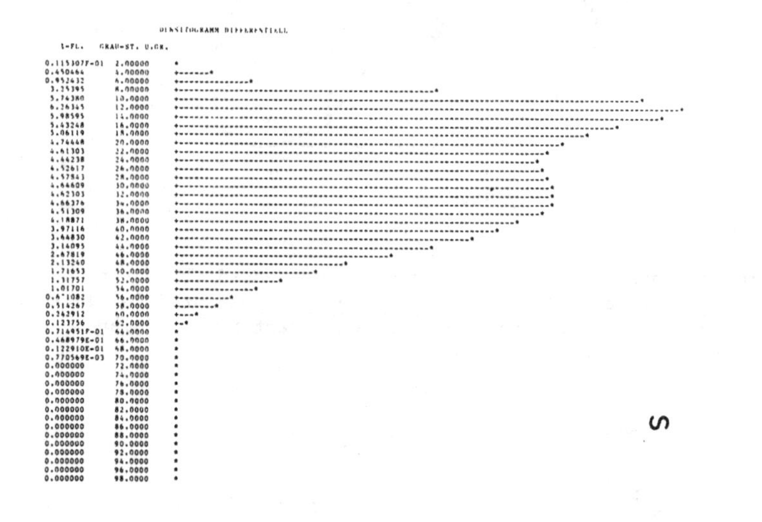

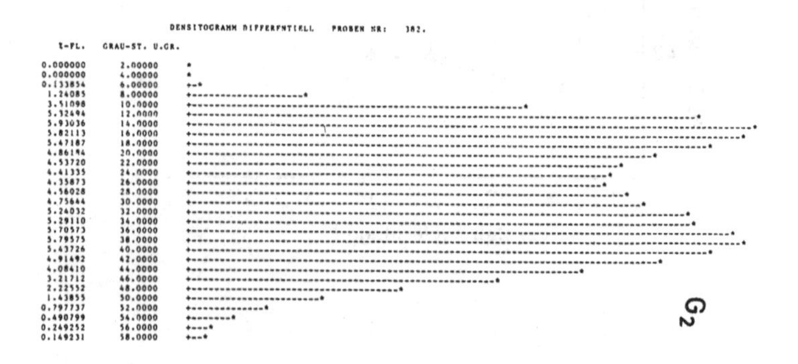

during G1 (Nicolini, 1979). However, there is no complete agreement between different laboratories (see, for instance, Rao & Hanks, 1980). A similar situation exists in plants where the "dispersion phase" (i.e. a stage during which heterochromatin decondenses) has not been unequivocally located to a particular stage of the cell cycle (Nagl, 1970a, 1977; Barlow, 1976).

Recently, sophisticated techniques were elaborated for the analysis of structural changes during the cell cycle using, for instance, monoclonal antibodies (Bhorjee, Barclay, Wedrychowski & Smith, 1983), DNase I sensitivity (Prentice, Tobey & Gurley, 1983), high-resolution TV-scanning and multivariant analysis (Abmayr, Giaretti, Gais & Dörmer, 1982).

Regulatory aspects

Although it is well established that the progress through the cell cycle depends on sufficient supply of nutritive materials, oxygen, and on certain hormones, and the synthesis of RNA and proteins, these factors apparently do not represent the basis of control. This was clearly expressed by Nicolini (1979): "The critical role in the control of cell proliferation ...could be linked not to the activation of specific genes (with most parts of the genome being permanently switched off) but to the overall periodic geometry of the genome at the three interrelated levels (of condensation), as determined by the chemical-electrostatic environment, mostly in terms of proteins (especially of histone H1 and non-histones, with the octamers as back-bones) or ions (as Ca^{++} and Mg^{++}) modifications, and partly in terms of the viscosity-diffusion properties *per se* of the highly condensed chromatin-DNA and of its surrounding aqueous medium".

Perhaps the regulation of the cell cycle can be considered at the physical level. Syngergetics describes the cooperation of the individual parts of a system that produces macroscopic spatial, temporal or functional structures in a self-organizing way. In synergetic terms, and in terms of non-linear thermodynamics of open systems far away from equilibrium (as characteristic for living systems), DNA/chromatin and their environment represent a co-operative system displaying properties such as self-organization and auto-catalysis. From this physical point of view a model of the cell cycle was proposed recently (Nagl & Popp, 1983) based on the electronic properties of DNA, ultraweak photon emission from biological systems, and thermodynamic and quantum theoretical consequences which can be drawn from experimental results (for details see Eisinger, 1968; Duchense, 1973).

The central property of the model can be seen in the evidence of excimer (excited dimers) and exciplex (excited complexes) formation of polynucleotides at room temperature within the range of the lowest triplet states of DNA. Excited complexes are molecules which join energetically excited electrons, and which emit the photons when the electrons fall back to a lower energy level. The latter event results in chemoluminescence and bioluminescence of various degrees, and is quite different from Van de Waal's interactions. Statistics concerning the emission of photons from living cells and seedlings, the long-distance effectivity of the emitted photons, and quantum-mechanical calculations are all consistent with the assumption that at least part of the radiation is coherent. (Li, Popp, Nagl & Klima, 1983; Popp _et al._, 1984). Coherence means that the quanta possess the same energy and frequency, as in a laser. Coherent light emission is, therefore, able to excite molecules (e.g. enzymes) in a highly specific way. Fig. 7 shows the events which play an elemental role for the cell cycle according to this model. There are two antagonistic external factors, namely energy loss due to chaotic influences, and energy supply due to "pumping" (nutrition).

Fig. 7. Control circuit between order and chaos, exciplex formation in DNA and decay, and the cell cycle according to the physical model of Nagl and Popp (1983).

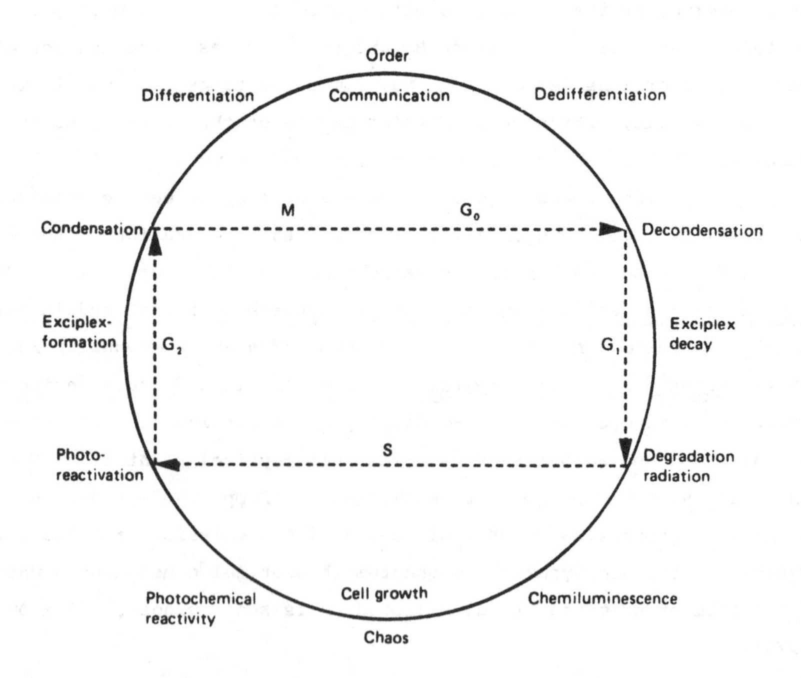

Besides these external events we also find two internal antagonistic factors, namely photon absorption (the photochemical potential) of photon stores as a measure of the photon uptake from the environment (e.g. glycolysis), and the resonator value (Q), which describes the capacity of photon storage (e.g. by excimer formation). This Q value is a measure of the self-information of the biological structure. The second term of importance, the photochemical potential, probably has its very origin in the Bose-Einstein condensation, which is expressed here through the attractive potential of excited base pairs and possibly other molecular DNA fragments. It is, therefore, a quantity which is accessible to measurements and quantum theoretical calculations. Such measurements and computations are now in progress, in order to analyze the role of interactions between chromatin structure and photon absorption and emission, respectively, in the control of the cell cycle.

CONCLUSIONS

Changes in chromatin structure during the cell cycle evidently occur in all organisms. They can be objectively described and digitized at the light and electron microscope level. They can be envisaged as the central aspect of the self-regulation of the cell cycle as they are the cause, and at the same time the consequence of the progression through the mitotic cycle. The ultimate basis of this cybernetic system is thought to lie in the physical properties of DNA and chromatin.

REFERENCES

Abmayr, W., Giaretti, W., Gais, P. & Dörmer, P. (1982). Discrimination of G1, S, and G2 cells using high-resolution TV-scanning and multivariate analysis methods. Cytometry, 2, 316-26.

Altmann, H.-W. (1966). Der Zellersatz, insbesonders an parenchymatösen Organen. Verhandlungen Deutscher Gesellschaft Pathologie 50, 15-51.

Barlow, P.W. (1976). The relationship of the dispersion phase of chromocentric nuclei in the mitotic cycle to DNA synthesis. Protoplasma, 90, 381-91.

Bhorjee, J.S., Barclay, S.L., Wedrychowski, A. & Smith, A.M. (1983). Monoclonal antibodies specific for tight-binding human chromatin antigens reveal structural rearrangements within the nucleus during the cell cycle. Journal of Cell Biology, 97, 389-96.

De la Torre, C., Sacristán-Gárate, A. & Navarrete, M.H. (1975). Structural changes in chromatin during interphase. Chromosoma, 51, 183-98.

Duchesne, J. (ed.), (1973). Physico-Chemical Properties of Nucleic Acids. New York: Academic Press.

Eisinger, J. (1968). Excited states of DNA. Science, 161, 1311-19.

Galbraith, D.W., Harkins, K.R., Maddox, J.M., Ayres, N.M., Sharma, D.P. & Firoozabady, E. (1983). Rapid flow cytometric analysis of the cell cycle in intact plant tissues. Science, 220, 1049-51.

Heller, F.O. (1973). DNS-Bestimmung an Keimwurzeln von Vicia faba L. mit Hilfe der Impulscytophotometrie. Bericht der Deutschen botanischen Gesellschaft, 86, 437-41.

Kendall, F., Swenson, R., Borun, T., Rowinski, J. & Nicolini, C. (1977). Nuclear morphology during the cell cycle. Science, 196, 1106-9.

Klueva, T.S., Onishchenko, G.E., Poliakov, V. Yu. & Chenstow, Yu.S. (1974). Ultrastructure of the cell interphase nuclei of some plants in different periods of mitotic cycle. Tsitologiya, 16, 1465-9.

Lafontaine, J.-G. & Lord, A. (1969). Organisation of nuclear structures in mitotic cells. In Handbook of Molecular Cytology, ed. A. Lima-de-Faria, pp.381-411. Amsterdam & London: North Holland.

Lafontaine, J.-G. & Lord, A. (1974). An ultrastructural and radioautographic study of the evolution of the interphase nucleus in plant meristematic cells (Allium porrum). Journal of Cell Science, 14, 263-87.

Li, K.H., Popp, F.A., Nagl, W. & Klima, H. (1983). Indication of optical coherence in biological systems and its possible significance. In Coherent Excitations in Biological Systems, eds. H. Fröhlich & F. Kremer, pp. 117-22. Berlin: Springer-Verlag.

Mazia, D. (1963). Synthetic activities leading to mitosis. Journal of Cellular and Comparative Physiology, 62, Suppl. 1, 123-40.

Mitchell, J.P., Cohn, N.S. & Van der Ploeg, M. (1983). Quantitative changes in the degree of chromatin condensation during the cell cycle in differentiating Pisum sativum vascular tissue. Histochemistry, 78, 101-9.

Moyne, G., Bertaux, O. & Puvion, E. (1975). The nucleus of Euglena. I. An ultracytochemical study of the nucleic acids and nucleo-proteins of synchronized Euglena gracilis Z. Journal of Ultrastructural Research, 52, 362-76.

Müller, H.-A. (1966). Die Chromozentren in den Leberzellkernen der Maus unter normalen und pathologischen Bedingungen. Ergebnisse der allgemeinen Pathologie und pathologischen, 47, 143-85.

Nagl, W. (1968). Der mitotische und endomitotische Zellzyklus bei Allium carinatum. I. Struktur, Volumen und DNA-Gehalt der Kerne. Osterreichische botanische Zeitschrift, 115, 322-53.

Nagl, W. (1970a). The mitotic and endomitotic nuclear cycle in Allium carinatum. II. Relations between DNA replication and chromatin structure. Caryologia, 23, 71-8.

Nagl, W. (1970b). Correlation of chromatin structure and interphase stage in nuclei of Allium flavum. Cytobiologie, 1, 395-8.

Nagl, W. (1972). Evidence of DNA amplification in the orchid Cymbidium in vitro. Cytobiologie, 5, 145-54.

Nagl, W. (1977). Nuclear structures during cell cycles. In Mechanisms and Control of Cell Division, eds. T.L. Rost & E.M.Gifford, Jr., pp.147-93. Stroudsberg, Pa: Dowden, Hutchinson & Ross.

Nagl, W. (1982). Condensed chromatin: species-specificity, tissue-specificity, and cell cycle-specificity, as monitored by scanning cytophotometry. In Cell Growth, ed. C. Nicolini, pp. 171-218. New York: Plenum.

Nagl, W. (1983). Computer-aided identification and digitation of cell cycle-specific chromatin ultrastructure. Journal of Interdisciplinary Cycle Research, 14, 53-61.

Nagl, W. & Popp, F.A. (1983). A physical (electromagnetic) model of differentiation. I. Basic considerations. Cytobios, 37, 45-62.

Nagl, W. & Rucker, W. (1976). Effects of phytohormones on thermal denaturation profiles of Cymbidium DNA: indication of differential DNA replication. Nucleic Acids Research, 3, 2033-9.

Nagl, W., Jeanjour, M., Kling, H., Kühner, S., Michels, I., Müller, T. & Stein, B. (1983). Genome and chromatin organization in higher plants. Biologische Zentralblatt, 102, 129-48.

Nicolini, C. (1979). Chromatin structure, from Angstrom to micron levels, and its relationship to mammalian cell proliferation. In Cell Growth, ed. C. Nicolini, pp. 613-66. New York: Plenum.

Nicolini, C. (1980). Nuclear morphology, quinternary chromatin structure and cell growth. Journal of Submicroscopy and Cytology 12, 475-505.

Nicolini, C., Ajiro, K., Borun, T.W. & Baserga, R. (1975). Chromatin changes during the cell cycle of HeLa cells. Journal of Biological Chemistry, 250, 3381-5.

Nicolini, C., Kendall, F. & Giaretti, W. (1977). Objective identification of cell cycle phases and subphases by automated image analysis. Biophysics Journal, 19, 163-76.

Olszewska, M.J. & Marciniak, K. (1977). The role of histones in the restriction of chromatin activity in successive stages of development of the antheridial filaments of Chara vulgaris L. Folia Histochemica et Cytochemica, 15, 109-19.

Popp, F.A., Nagl, W., Li, K.H., Scholz, W., Weingärtner, O. & Wolf, R. (1984). Biophoton emission. New evidence for coherence and DNA as source. Cell Biophysics (in press).

Prentice, D.A., Tobey, R.A. & Gurley, L.R. (1983). DNase I and cellular factors that affect chromatin structure. Biochimica et Biophysica Acta, 741, 288-96.

Rao, P.N. & Hanks, S.K. (1980). Chromatin structure during the pre-replicative phases in the life cycle of mammalian cells. Cell Biophysics, 2, 327-37.

Sawicki, W., Rowinski, J. & Abramczuk, J. (1974). Image analysis of chromatin in cells of preimplantation mouse embryos. Journal of Cell Biology, 63, 227-33.

Schweizer, D. & Nagl, W. (1976). Heterochromatin diversity in Cymbidium, and its relationship to differential DNA replication. Experimental Cell Research, 98, 411-23.

Setterfield, G., Hall, R., Bladon, T., Little, J. & Kaplan, J.G. (1983). Changes in structure and composition of lymphocyte nuclei during mitogenetic stimulation. Journal of Ultrastructural Research, 82, 264-82.

Troyan, V., Kolesnikov, V., Kalinin, F., Zelenin, A. & Ilchenko, L. (1984). The periodicity of the changes in the rate of RNA synthesis, endogenous RNA polymerase activity and DNA availability to acridine orange during the first cell cycle in the root meristematic cells of germinating Pisum sativum L. Plant Science Letters, 33, 213-9.

THE CYTOSKELETON AND THE PLANT CELL CYCLE

L. Clayton

INTRODUCTION

The plant cytoskeleton has a number of features in common with those of animal cells in that the major components, such as microtubules and microfilaments, appear ultrastructurally identical and the basic molecular building-blocks (e.g. tubulin and actin) are very similar to their animal counterparts. The ways in which these components are arranged and participate in the cell cycle, however, appear very different and reflect different strategies of morphogenesis; in animals, migration of cells plays a large part in the formation of tissues, whereas in plants, which have cell walls that preclude migration, morphogenesis is primarily achieved by the precise manner in which cells co-ordinate division and elongation. The way in which the cytoskeleton functions, and the rearrangements which take place during the plant cell division cycle are fundamental to the subsequent tissue patterning and morphogenesis.

The involvement of the cytoskeleton in various aspects of plant development has recently been the subject of a book which reviews the field in depth (Lloyd, 1982). This chapter will deal mainly with the cytoskeleton in dividing higher plant cells in an attempt to give a broad view of established work and to highlight the directions in which current research is proceeding.

The microtubule cycle

Microtubules are perhaps the best documented component of the plant cytoskeleton and play a major role throughout the cell cycle, influencing patterns of wall growth during interphase and participating in both nuclear division (by means of mitotic/meiotic spindles) and cell separation (development of new cross walls).

Microtubules were first described in plant cells by Ledbetter & Porter (1963) as a result of the advent of the use of aldehyde fixatives in the preparation of specimens for electron microscopy. Subsequent research by a number of workers enabled Ledbetter (1967) to describe, for the first time, the arrangement of microtubules throughout the cell division cycle of higher plants in his review of much of the early ultrastructural work on plant cell division.

Throughout interphase, in meristematic cells, microtubules occupy the cortex of the cell, and are arranged circumferentially around the cell axis in what are often referred to as 'hoops', although more recent evidence suggests that in some, if not all cells, microtubules, or bundles of microtubules, describe a helical pattern around the cell (Lloyd, 1983). The pattern of microtubules often parallels the pattern of cellulose microfibrils in the wall (Ledbetter & Porter, 1963; Preston, 1974; Robinson & Quader, 1982). There seems little doubt that microtubules, and possibly other components of the cytoskeleton, influence the orientation of cellulose although the molecular and cellular mechanism by which this takes place is at present the subject of speculation (Heath & Seagull, 1982; Lloyd, 1984a).

There is ample evidence that cortical microtubules are attached to the cell membrane: links between these two structures may be seen in electron micrographs (Franke *et al.*, 1972). Microtubules remain attached to the exposed cytoplasmic face of burst protoplasts (Marchant, 1978; Van der Valk, Rennie, Connolly & Fowke, 1980) and it seems that such links can survive artificially induced shape changes e.g. during the conversion of elongated cells to round protoplasts (Lloyd *et al.*, 1980). It is obvious from examination of successive and variably orientated layers of cell wall, that for parallelism to persist throughout many phases of wall deposition, the organisation of microtubules must be flexible, implying that the interphase array must be a dynamic structure. Ideas as to how, or even whether this is achieved abound (Lloyd, 1984b, for a review).

As the plant prepares for division in organised tissue, the interphase array gives way to an equatorial band of closely-packed microtubules in the cortex, which encircles the nucleus. This structure was first described by Pickett-Heaps & Northcote (1966) in wheat (*Triticum aestivum*) meristems and developing stomatal complexes in leaves, and is commonly referred to as the pre-prophase band. Remarkably, this band predicts the plane of the subsequent cell division, i.e. the cortical sites at which the cell will place its new dividing wall, with great accuracy. This ability to predict even asymmmetric divisions is illustrated by the study of Pickett-Heaps & Northcote (1966). Gunning, Hardham & Hughes (1978a) also showed that this was the case for every single division in the root of the water fern, Azolla. Moreover, this is also true when cells are stimulated to divide following wounding (Hardham & McCully, 1982; Venverloo, Hovenkamp, Weeda & Libbenga, 1980) i.e. in divisions which are not a normal part of the developmental programme.

In tissues where cells divide in a disorganised or random manner, such as callus, suspension culture cells or dividing protoplasts (Fowke, Boch-Hansen, Constabel & Gamborg, 1974; Fowke & Gamborg, 1980), pre-prophase bands are not evident, therefore it is tempting to correlate lack of morphogenetic order with lack of pre-prophase bands (Gunning, 1982).

Recent work using immunofluorescence localisation of microtubules (see later) in suspension cultured cells (Simmonds, Setterfield & Brown, 1983) suggests that pre-prophase bands may in fact be present but are less compact and well ordered, and of short duration in the cell cycle. These may be missed using conventional electron microscopy. In contrast, however, there are a number of examples of divisions which are precisely aligned without being preceded by a pre-prophase band e.g. most algae (Pickett-Heaps, 1978), some tissues of moss and ferns, and cellularising endosperm tissue. The question, therefore, as to whether the pre-prophase band influences, as opposed merely to predicting, division either by positioning the nucleus or polarising the spindle and/or cytokinetic apparatus, is problematical. Pickett-Heaps' (1974) view still seems to hold: the pre-prophase band can indeed 'predict' or anticipate the plane of division but its ability to exert influence is not proven. The pre-prophase band is usually completely dispersed at the onset of prophase proper.

The mechanism of mitosis in plant tissue has been studied most thoroughly in the liquid endosperm of Haemanthus katherinae (Bajer & Molè-Bajer, 1972; Jensen, 1982). This is a somewhat atypical plant tissue being a mainly wall-less triploid syncytium, but has the advantage of being very amenable to experimental manipulation and investigation using a wide variety of optical and electron microscopic techniques. Moreover mitosis in Haemanthus demonstrates one of the most conspicuous features of higher plant mitosis in general, namely that centrioles do not feature at the spindle poles (Pickett-Heaps, 1969). This is the case for all higher plant cells; only in some algae and certain tissues of ferns and mosses are centrioles present at the spindle poles, usually synthesised de novo in cells where they will eventually form the basal bodies of flagella e.g. in the fern Masilea (Hepler, 1976). In other respects plant mitosis, as studied in

Haemanthus, progresses in a similar manner to animal cells and is elegantly described in great detail by Bajer & Molè-Bajer (1981).

After the mitotic spindle has separated the chromosomes, cytokinesis occurs by the deposition of a new cell wall (the cell plate) between the daughter nuclei, beginning at telophase. This is accompanied by the appearance of a new array of microtubules, which is unique to higher plant cells, called the phragmoplast (Hepler & Newcomb, 1967; Hepler & Jackson, 1968). This is composed of two sets of interdigitating microtubules which appear in the mid-zone of the telophase spindle, and is thought to be concerned with directing vesicles to the site of cell plate formation. As the cell plate spreads outwards, centrifugally, to join the existing side walls of the cell, the phragmoplast expands outwards concomitantly, often far beyond the confines of the preceding spindle. Indeed the growing cell plate and phragmoplast may undergo extensive re-orientation during its growth in order to contact the side walls at the site 'predicted' by the pre-prophase band (Palevitz & Hepler, 1974; Gunning, 1982).

The four basic types of microtubule organisation described above constitute the typical microtubule cycle in dividing cells of higher plants. Most of this information has been obtained from conventional light microscopy and electron microscopic ultrastructural data. Recently techniques, such as indirect immunofluorescent localisation of cellular components, which have proved so fruitful in the investivation of animal cytoskeletons (Aubin, 1981), have been applied to plant cells. In general plant cells possess a significant barrier to permeation by antibodies in the shape of the cell wall; consequently the initial immunofluorescence studies of microtubules in plant cells were conducted using wall-less endosperm cells (Franks *et al*., 1977) in which the mitotic apparatus could be visualised. Subsequently the cortical interphase array was demonstrated by immunofluorescence of cellulase-treated suspension cultured cells (Lloyd *et al*., 1979).

A method enabling immunofluorescence studies to be applied to cells in organised tissues was devised by Wick *et al*., (1981) who isolated pre-fixed, permeabilised cells from root meristem tissue in which microtubules could be stained at all stages of the cell cycle, in cells which had retained their native shape. In such a way microtubule arrays were displayed for the first time in their entirety, in particular the variety of forms taken by the pre-prophase band (Wick & Duniec, 1983), an aspect of microtubule arrangement not readily apparent from conventional ultrastructural studies.

Many cytoskeletal proteins, especially tubulin, the protein constituent of microtubules, are highly conserved in evolutionary terms (Little, Krauhs & Postingl, 1981), the β subunit of the αβ heterodimer being the most similar throughout very diverse phyla: this property enables immunocytochemical techniques to be applied to plants in that antibodies raised against animals or microbial tubulins recognise plant microtubules.

Fig. 1 illustrates the microtubule cycle in onion (*Allium cepa* root meristem cells by immunofluorescent localisation of microtubules using an antibody against tubulin. One of the major advantages of this technique is that it permits the inspection of large populations of cells consisting of representatives from all stages of the cell division cycle. Such preparations may be used to assess the effects of various agents which disrupt the cytoskeleton and perturb the cell cycle of plants. Plant microtubules appear resistant to depolymerisation by conventional anti-microtubule agents such as colcemid and colchicine (Hart & Sabnis, 1976), requiring up to one thousand times the concentration effective upon animal cells. Two particular drugs, however, exert cell-cycle-specific effects upon microtubule organisation: the herbicide isopropyl N- (3-chlorophenyl) carbamate (CIPC) and the antifungal antibiotic griseofulvin disrupt spindle organisation producing tri- and often multi-polar spindles. These result in separation of chromosomes to form abnormal numbers of "nuclei" which are ultimately partitioned by the development of branched phragmoplasts and tortuous cell plates (see Fig. 2). In contrast, the interphase cortical arrays and pre-prophase bands are unaffected both in appearance and frequency. These results suggest that the predictive capacity of the pre-prophase band may be in some way uncoupled from cytokinesis (Clayton & Lloyd, 1984).

The precise mode of action of these drugs is as yet unclear, although much evidence from a variety of animal, plant and microbial systems (Gull & Trinci, 1973; Oliver, Krawiec & Berlin, 1978; Coss & Pickett-Heaps, 1974) implies that they act in some manner on the microtubule organising centres (MTOCs), of the spindle particularly, maybe disturbing their microtubule organising capacity and/or duplication.

Microfilaments

Microtubules play a major cytoskeletal role at all stages of the plant cell cycle, and not surprisingly most attention has been concentrated upon them. Consequently relatively little is known about the contribution of

Fig. 1. The microtubule cycle in higher plant meristematic cells: microtubules are visualised by immunofluorescence microscopy in root-tip cells of Allium cepa. Bar = 10 μM.
(a) Interphase cortical array in which microtubules are closely associated with the cell surface.
(b) Pre-prophase:- the interphase array gives way to a band of microtubules, in the cortex, which surrounds the nucleus and predicts the division site. Microtubules are also present at the nuclear surface, roughly focused at each pole.
(c) Prophase:- the pre-prophase band has disappeared and the spindle is beginning to form.
(d) Metaphase.
(e) Anaphase.
(f) Telophase:- phragmoplast microtubules are present in the midzone, where the cell plate is forming.
(g) Late telophase/Early interphase:- remnants of the phragmoplast are visible on one side of the cell plate, and a new interphase array is beginning to form. Microtubules appear to radiate from the nucleus.
(h) Interphase:- the cortical array is re-established.

other structural components to the internal architecture of plant cells.

In animal cells and other motile, wall-less cells, actin is a major component, present as networks or bundles of microfilaments. In higher plant cells, which are non-motile, microfilament bundles have been demonstrated in EM sections of vascular tissue and other cells where vigorous cytoplasmic streaming is known to occur (Williamson, 1980; Jackson, 1982). Such bundles have been identified as actin by decoration with heavy meromysin in extracts of Characean algae. The presence of a protein resembling animal actin has been confirmed by immunological methods (Metcalf, Szabo, Schubert & Wang, 1980), and recently by purification and assembly in vitro of actin filaments from plant tissue (Vahey, Titus, Trautwein & Scordilis, 1982); moreover actin-coding DNA sequences have been detected in soybean (Glycine max) using heterologous cDNA probes (Nagao, Shah, Eckenrode & Meagher, 1981).

Despite this work however, the intracellular distribution of actin microfilaments in meristematic cells has not been investigated. The technical advance of obtaining intact permeabilised cells from dividing tissue has enabled such an investigation using fluorescent cytochemical methods. Fluorescently labelled derivatives of phallotoxins from the fungus Amanita phalloides may be used specifically to localise filamentous actin in tissues of animal cells (Wulf et al., 1979), fungi (Hoch & Staples, 1983; Kilmartin & Adams, 1984) and the vascular tissues of plants (Persecreta, Carley, Webb & Parthasarathy, 1982) and such an approach has been used to

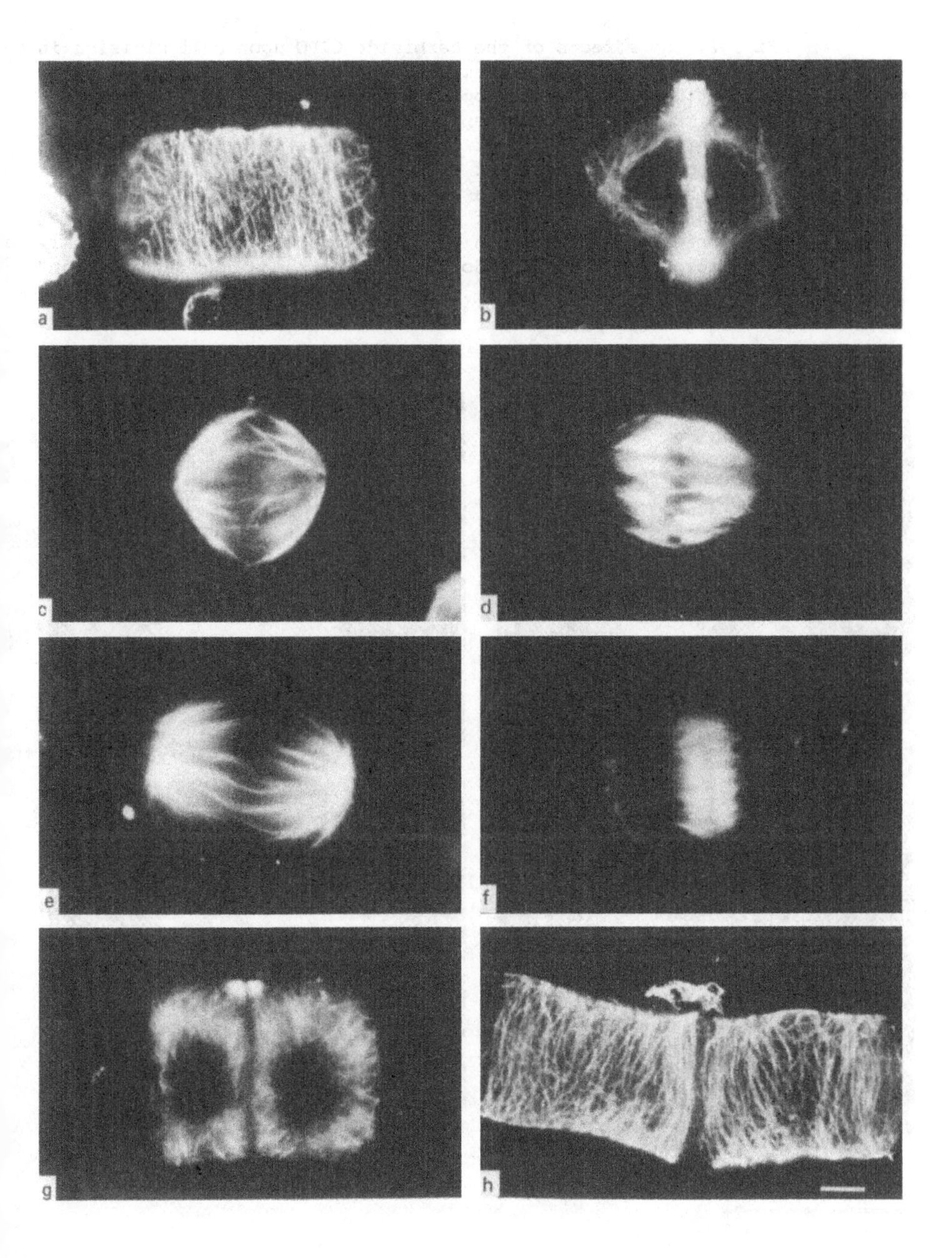
a
b
c
d
e
f
g
h

Fig. 2. The effects of the herbicide CIPC upon cell division in _Allium_. Microtubules are visualised by immunofluorescence, and chromatin stained with propidium iodide. Bar = 10 μM.

(a) A tripolar anaphase spindle.

(b) Chromatin staining of the cell in (a), showing chromosomes separating to each of the three poles.

(c) A branched phragmoplast.

(d) Chromatin staining of (c), showing three "nuclei". The pre-prophase bands in CIPC-treated cells show no abnormal morphology and thus do not predict these unusual division planes.

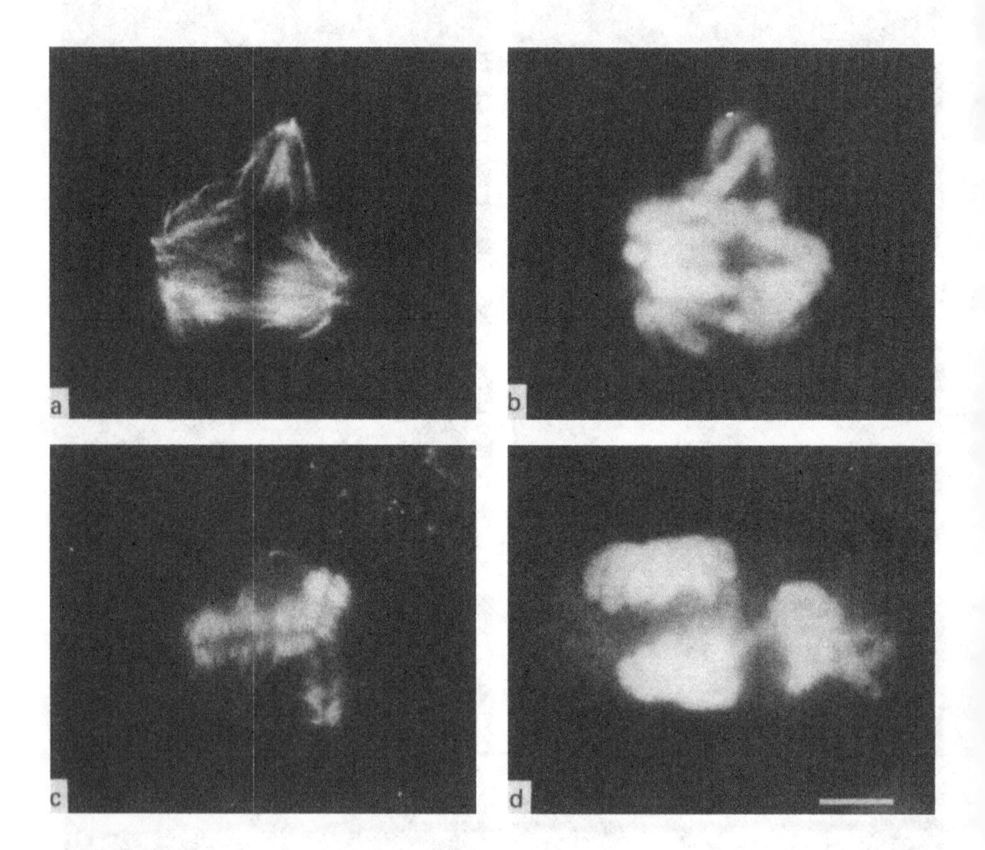

visualise F-actin in preparations of root meristem tissue (Clayton & Lloyd, 1984).

Bundles of microfilaments are present in the cytoplasm of elongated cells, frequently in close association with the nucleus and often spanning the length of the cell, running at right angles to the cortical microtubule array (Fig. 3). These bundles are not present during mitosis, and microfilaments cannot be detected in the spindle using fluorescent phalloidin staining. However, during cytokinesis F-actin appears to be present in the phragmoplast and co-localised with microtubules at all stages of cell plate formation (Fig. 3).

These images suggest that microfilaments may not be involved in chromosome movement in mitosis but may play a part, along with microtubules, in the movement and/or direction of vesicles to the growing cell plate. Microfilaments could also participate in nuclear anchoring and movement as well as in cytoplasmic streaming during interphase.

New cytoskeletal components

Many types of animal cells possess a third system of filaments in the cytoplasm - namely intermediate or 10 nm filaments, being intermediate in diameter between 6 nm actin microfilaments and 25 nm microtubules (Lazarides, 1980). Such a system has not, as yet, been detected or indeed been seriously searched for in plant cells or tissues. Bundles of fibrillar material have been observed in thin sections of carrot (*Daucus carota*) suspension cells (Wilson, Israel & Steward, 1974) and similar bundles have been shown to be a prominent feature or detergent-extracted suspension-cultured cells of a number of plant species (Powell, Peace, Slabas & Lloyd, 1982). They appear not to be composed of actin or tubulin but the precise biochemical nature and formation of such fibrillar bundles has yet to be determined.

Immunofluorescence techniques now enable whole isolated plant cells to be probed for other cytoskeletal proteins and structures utilising antibodies raised against animal proteins. Such an approach presupposes and relies upon a certain degree of similarity between animal and plant proteins, an assumption which appears to be justified with respect to tubulin which is a highly conserved protein. This is well illustrated by the fact that a monoclonal antibody against yeast (*Saccharomyces cerevisiae*) tubulin (Kilmartin, Wright & Milstein, 1982) cross-reacts with plant tubulins as well as those of a very wide range of other animal and microbial sources. Using suitable

Fig. 3. The distribution of F-actin during the cell cycle in Allium meristematic cells, stained with a fluorescent derivative of phalloidin - a specific reagent for assembled actin. Bar = 10 µM.
(a) Bundles of F-actin in elongated interphase cells. These bundles follow the long axis of the cell, often in close proximity to the nucleus. They may be involved in cytoplasmic streaming.
(b) Phalloidin staining of actin in the phragmoplast.
(c) & (d) A cell double stained for actin (c), and microtubules (d), showing the co-distribution of microtubules and actin in the phragmoplast.
No phalloidin-specific staining is seen during pre-prophase mitosis.

antibodies, this approach may reasonably be expected to be informative when applied to plants. It is to be hoped that the use of immunocytochemistry, coupled with monoclonal antibody technology, will lead to the discovery of further cytoskeletal proteins in plants, as has been the case for animal cells (for review see Birchmeir, 1984).

Control of the cytoskeleton in the cell cycle

We have seen that the cytoskeleton of meristematic cells undergoes a number of rearrangements associated with cell cycle events, parti cularly preparation for mitosis and cytokinesis followed by the re-establishment of a new interphase array. How these re-arrangements are controlled and integrated with other cell cycle processes is not known but is an area of active research in a variety of systems.

In all cells, microtubules are organised in a site-specific manner; the areas of the cell responsible for microtubule nucleation and organisation are known as microtubule organising centres (MTOCs)(Pickett-Heaps, 1969; Brown, Sterns & McRae, 1982). These areas may take a variety of forms in different types of cells, ranging from the centrosome of most mammalian cells (McIntosh, 1983) to the flagellar rootlet structures in the algae *Chlamydomonas* and *Polytomella* (Brown *et al*., 1982) but in all cases investigated the material in which the microtubules are embedded, and which appears to nucleate the microtubules, is structurally amorphous and electron-dense in conventionally stained EM thin sections.

In higher plants very little is known about the location and behaviour of MTOCs. Amorphous material similar in appearance to the pericentriolar MTOC material of animal cells can be observed at the poles of

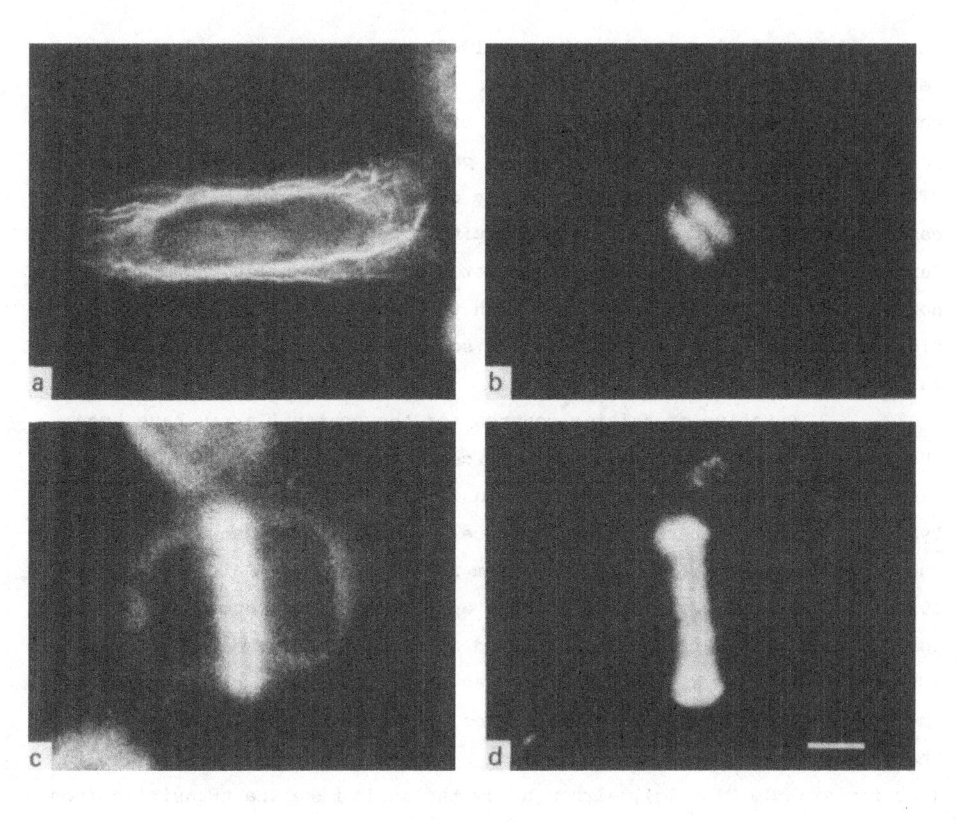
a
b
c
d

higher plant spindles (Pickett-Heaps, 1969) but its behaviour during the rest of the cell cycle is not known: for instance is this the same material which participates in the organisation of the phragmoplast and/or the interphase cortical MT arrays? Animal cells possess convenient markers for the MTOC, the centrioles, which are usually closely associated with pericentriolar MTOC material both at the spindle poles during mitosis, and during interphase (McIntosh, 1983). Most higher plant cells, however, do not contain centrioles, consistent with the lack of ciliated/flagellated tissues or gametes in land plants, and so MTOC behaviour cannot be followed using such a marker.

Putative organising centres for the cortical array have been observed in thick sections of root tip cells examined in the high voltage electron microscope. They appear as clusters of vesicles associated with typical amorphous material lining the cell edges, which seem to serve as foci for microtubules (Gunning, Hardham & Hughes, 1978b; Palevitz, 1981, 1982). It has been proposed that these apparent cortical nucleating sites for the interphase arrays are inherited at cytokinesis by daughter cells (Gunning, 1980). An alternative view may be proposed from observations of immunofluorescent images of microtubules. In late telophase/early interphase, cells often show microtubules apparently radiating from the nucleus (see for example Fig. 1g), although how the cells make the transition from this to a complete interphase array is unclear. The nuclear envelope has been suggested as a possible nucleating site for microtubules in _Haemanthus_ endosperm cells (Schmit, Vantard, de Mey & Lambert, 1983); in fact, close association of MTOCs with the nucleus is common in many animal and fungal cells e.g. in yeast and many other fungi the spindle pole MTOCs are integral with the nuclear envelope (Heath, 1978). Moreover, MTOCs may be co-isolated with nuclei from _Dictyostelium_ (Kuriyama, Sato, Fukui & Nishibayashi, 1982), _Physarum_ (Roobol, Havercroft & Gull, 1982) and cultured animal cells (Kuriyama & Borisy, 1981). Such a possibility in plant cells has not been adequately explored.

Even more enigmatic is the origin of the phragmoplast microtubules; osmiophilic amorphous material similar to that seen at spindle poles is frequently observed at the mid-zone of the phragmoplast where vesicles of cell wall material are fusing (Hepler & Jackson, 1968) and this material has been assumed to represent the nucleation site of the phragmoplast microtubules. Recent studies on the polarity of microtubules in this array have, however, cast doubts on this assumption (Euteneuer & McIntosh,

1980).

Microtubules are essentially polar structures, having a fast-growing ("plus") end and a slow-growing ("minus") end (Margolis & Wilson, 1978; Bergen & Borisy, 1980). In vivo, most microtubules are associated with an MTOC, and for most of the microtubule-containing structures examined in cells it is the "plus" end of microtubules which is distal to the MTOC. Techniques are available by which microtubules may be "decorated", either with purified dynein ATP-ase arms (Haimo, Telzer & Rosenbaum, 1979) or with hooks formed by the polymerisation of exogenous tubulin under bizarre buffer conditions (Heidemann & McIntosh, 1980), in order to reveal their polarity, in much the same way that heavy meromyosin decorates actin filaments. Using these techniques to examine Haemanthus phragmoplasts, it appears that the "plus" ends of these microtubules are directed towards the mid-zone, suggesting either that the MTOC for this array is not located in the mid-zone, or that these microtubules are nucleated in the mid-zone with their "minus" ends distal to the MTOC. This question has yet to be resolved.

The mechanism(s) by which MTOCs control the spatial and temporal re-arrangements of microtubules during the cell cycle are not known although evidence is accumulating on the factors which may influence microtubule assembly in vivo. Most attention has been devoted to the role of calcium which is well known to affect microtubules in vitro. Changes in free calcium levels have been demonstrated during mitosis in Haemanthus endosperm using fluorescent probes for calcium (Wolniak, Hepler & Jackson, 1983); moreover an extensive membraneous system at the spindle poles has been implicated in calcium sequestration (Wick & Hepler, 1980). In addition, the calcium-binding protein calmodulin has been detected in association with spindles both in animal cells (Zavortink, Welsh & McIntosh, 1983) and also recently in Haemanthus, using immuno-fluorescence microscopy (Lambert, Vantard, Van Eldick & De Mey, 1982) although there is as yet no evidence to suggest its function (also see Trewavas, this volume).

A further level of cytoskeletal control may operate at the genetic level. In many species from animals to micro-organisms there are multiple copies of the genes for cytoskeletal proteins (Cowan & Dudley, 1984) and in some cases multiple protein species can be detected. Fulton & Simpson (1976) proposed that different tubulin sub-species may function in an organelle-specific manner and there is some evidence for this idea from work on Chlamydomonas and Polytomella (McKeithan, Lefebvre, Silflow & Rosenbaum, 1983) in which a specific α-tubulin species is assembled into flagella.

With respect to plants, it is known that soybean (*G. max*) DNA contains many actin-related sequences, as detected by use of heterologous cDNA probes, but it is not known how these are expressed. A similar situation exists with plant tubulin and although tubulin has been purified from plant tissue (Morejohn & Fosket, 1982), neither the number of protein species nor the number of genes is known. We are still some way from understanding the part played by gene expression in cytoskeletal control, and even in systems where more information is available (e.g. Cowan & Dudley, 1984 for review) the relationship between genetic complexity and microtubule function has yet to be determined.

CONCLUSIONS

The structural rearrangements of the plant cell cytoskeleton during the cell cycle have profound consequences with respect to subsequent morphogenesis; the control of wall growth and cell expansion together with the determination of division planes are the basis of plant morphogenesis, and the cytoskeleton is intimately involved in these processes at all stages of the cell cycle.

Although the basic cytoskeletal components of plants appear to be similar to their animal counterparts we know comparatively little about how they interact, what other components are involved and how the system is controlled. It is these questions to which plant cell biologists are turning their attention.

ACKNOWLEDGEMENTS

I am grateful to Drs C.W. Lloyd and K. Roberts for helpful discussion. I also thank the Royal Society and the John Innes Institute for support.

REFERENCES

Aubin, J.E. (1981). Immunofluorescence studies of cytoskeletal proteins during cell division. In *Mitosis/Cytokinesis*, ed. A.M. Zimmerman & A. Forer, pp. 211-44. New York: Academic Press.

Bajer, A.S. & Molè-Bajer, J. (1972). Spindle dynamics and chromosome movements. *International Review of Cytology*, supplement 3.

Bajer, A.S. & Molè-Bajer, J. (1981). Asters, poles and transport properties within spindle-like microtubule arrays. *Cold Spring Harbour Symposia on Quantitative Biology*, 46, 263-84.

Bergen, L.G. & Borisy, G.G. (1980). Head-to-tail polymerisation of microtubules *in vitro*. *Journal of Cell Biology*, 84, 141-50.

Brown, D.L., Sterns, M.E. & McRae, T.H. (1982). Microtubule organising centres. In The Cytoskeleton in Plant Growth and Development, ed. C.W. Lloyd, pp. 55-84. London: Academic Press.
Birchmeir, W. (1984). Cytoskeleton structure and function. Trends in Biochemical Sciences, 9, 192-4.
Clayton, L. & Lloyd, C.W. (1984). The relationship between the division plane and spindle geometry in Allium cells treated with CIPC and griseofulvin: an anti-tubulin study. European Journal of Cell Biology, 34, 248-53.
Coss, R.A. & Pickett-Heaps, J.D. (1974). The effects of isopropyl N-phenyl carbamate on the green alga Oedogonium cardiacum. I. Cell division. Journal of Cell Biology, 63, 84-97.
Cowan, N.J. & Dudley, L. (1984). Tubulin isotypes and the multigene tubulin families. International Review of Cytology, 85, 147-73.
Euteneuer, U. & McIntosh, J.R. (1980). Polarity of midbody and phragmoplast microtubules. Journal of Cell Biology, 87, 509-15.
Fowke, L.C., Boch-Hansen, C.W., Constabel, F. & Gamborg, O.L. (1974). A comparative study on the ultrastructure of cultured cells and protoplasts of soybean during cell division. Protoplasma, 81, 189-203.
Fowke, L.C. & Gamborg, O.L. (1980). Application of protoplasts to the study of plant cells. International Review of Cytology, 68, 9-51.
Franke, W.W., Herth, W., Van Der Woude, J. & Morré, D.J. (1972). Tubular and filamentous structures in pollen tubes; possible involvement as guide elements in protoplasmic streaming and vectorial migration of secretory vesicles. Planta, 105, 317-41.
Franke, W.W., Seib, E., Osborn, M., Weber, K., Herth, W. & Falk, H. (1977). Tubulin-containing structures in the anastral mitotic apparatus of endosperm of the plant Leucojum aestevum as revealed by immunofluorescence microscopy. Cytobiology, 5, 24-48.
Fulton, C. & Simpson, P.A. (1976). Selective synthesis and utilization of flagellar tubulin. The multi-tubulin hypothesis. In Cell Motility, ed. R. Goldman, T. Pollard, J. Rossenbaum, pp. 987-1005. New York: Cold Spring Harbor.
Gull, K. & Trinci, A.P.J. (1973). Griseofulvin inhibits fungal mitosis. Nature, 244, 292-4.
Gunning, B.E.S. (1980). Spatial and temporal regulation of nucleating sites for arrays of cortical microtubules in root tip cells of the water fern Azolla pinnata. European Journal of Cell Biology, 23, 53-65.
Gunning, B.E.S. (1982). The cytokinetic apparatus: its development and spatial regulation. In The Cytoskeleton in Plant Growth and Development, ed. C.W. Lloyd, pp. 229-94. London: Academic Press.
Gunning, B.E.S., Hardham, A.R. & Hughes, J.E. (1978a). Pre-prophase bands of microtubules in all categories of formative and proliferative cell division in Azolla roots. Planta, 143, 145-60.
Gunning, B.E.S., Hardham, A.R. & Hughes, J.E. (1978b). Evidence for initiation of microtubules in discrete regions of the cell cortex of Azolla root tip cells and an hypothesis on the development of cortical arrays of microtubules. Planta, 143, 161-79.
Haimo, L.T., Telzer, B.R. & Rosenbaum, J.L. (1979). Dynein binds to and crossbridges cytoplasmic microtubules. Proceedings of the National Academy of Sciences, USA. 76, 5759-63.
Hardham, A.R. & McCully, M.R. (1982). Reprogramming of cells following wounding in pea (Pisum sativum L.) roots. I. Cell division and differentiation of new vascular elements. Protoplasma , 112, 143-51.

Hart, J.W. & Sabnis, D.D. (1976). Colchicine and plant microtubules: a critical evaluation. Current Advances in Plant Science, 26, 1095-104.

Heath, I.B. (ed.) (1978). Nuclear Division in Fungi, New York & London: Academic Press.

Heath, I.B. & Seagull, R.W. (1982). Oriented cellulose fibrils and the cytoskeleton: a critical comparison of models. In: The Cytoskeleton in Plant Growth and Development, ed. C.W. Lloyd, pp. 163-84. London: Academic Press.

Heidemann, S.R. & McIntosh, J.R. (1980). Visualisation of the structural polarity of microtubules. Nature, 286, 517-9.

Hepler, P.K. (1976). The blepharoplast of Marsilea: its de novo formation and spindle association. Journal of Cell Science, 21, 361-90.

Hepler, P.K. & Jackson, W.T. (1968). Microtubules and early stages of cell plant formation in the endosperm of Haemanthus katherinae Baker. Journal of Cell Biology, 38, 437-46.

Hepler, P.K. & Newcomb, E.H. (1967). Fine structure of cell plate formation in the apical meristem of Phaseolus roots. Journal of Ultrastructural Research, 19, 498-513.

Hoch, H.C. & Staples, R.C. (1983). Visualisation of actin in situ by rhodamine-conjugated phalloidin in the fungus Uromyces phaseoli. European Journal of Cell Biology, 32, 52-8.

Jackson, W.T. (1982). Actomysin. In: The Cytoskeleton in Plant Growth and Development, ed. C.W. Lloyd, pp. 3-20. London: Academic Press.

Jensen, C.G. (1982). Dynamics of spindle microtubule organisation: kinetochore fibre microtubules of plant endosperm. Journal of Cell Biology, 92, 540-58.

Kilmartin, J.V. & Adams, A. (1984). Structural rearrangements of tubulin and actin during the cell cycle of the yeast Saccharomyces. Journal of Cell Biology, 98, 922-33.

Kilmartin, J.V., Wright, B. & Milstein, C. (1982). Rat monoclonal antitubulin antibodies derived by using a new non-secreting rat cell line. Journal of Cell Biology, 93, 576-82.

Kuriyama, R. & Borisy, G.G. (1981). Centriole cycle in Chinese hamster ovary cells as determined by whole-mount electron microscopy. Journal of Cell Biology, 91, 814-21.

Kuriyama, R., Sato, C., Fukui, Y. & Bishibayashi, S. (1982). In vitro nucleation of microtubules from microtubule organising centers prepared from cellular slime mould. Cell Motility, 2, 257-72.

Lambert, A.M., Vantard, M., Van Eldik, L. & de Mey, J. (1982). Immunolocalisation of calmodulin in higher plant endosperm cells during mitosis. Journal of Cell Biology, 97, 40a.

Lazarides, E. (1980). Intermediate filaments as mechanical integraters of cellular space. Nature, 283, 249-56.

Ledbetter, M.C. (1967). The disposition of microtubules in plant cells during interphase and mitosis. In: Formation and Fate of Cell Organelles, ed. K.B. Warren, pp. 65-70. London: Academic Press.

Ledbetter, M.C. & Porter, K.R. (1963). A microtubule in plant cell fine structure. Journal of Cell Biology, 19, 239-50.

Little, M., Krauhs, E. & Ponstingl, H. (1981). Tubulin sequence conservation. BioSystems, 14, 239-46.

Lloyd, C.W. (ed.) (1982). The Cytoskeleton in Plant Growth and Development. London: Academic Press.

Lloyd, C.W. (1983). Helical microtubular arrays in onion root hairs. Nature, 305, 311-3.

Lloyd, C.W. (1984a). Microtubules and the cellular morphogenesis of plants. In: Developmental Biology: a comprehensive synthesis, ed. Brower, in press. New York: Plenum.

Lloyd, C.W. (1984b). Towards a dynamic helical model for the influence of microtubules on wall patterns in plants. International Review of Cytology, 86, 1-51.

Lloyd, C.W., Slabas, A.R., Powell, A.J. & Lowe, S.B. (1980). Microtubules, protoplasts and plant cell shape. Planta, 147, 500-6.

Lloyd, C.W., Slabas, A.R., Powell, A.J., MacDonald, G. & Badley, R.A. (1979). Cytoplasmic microtubules of higher plant cells visualised with anti-tubulin antibodies. Nature, 278, 239-41.

Marchant, H.J. (1978). Microtubules associated with the plasma membrane isolated from protoplasts of the green alga Mougeotia. Experimental Cell Research, 115, 25-30.

Margolis, R.L. & Wilson, L. (1978). Opposite end assembly and disassembly of microtubules at steady state in vitro. Cell, 13, 1-8.

McIntosh, J.R. (1983). The centrosome as an organiser of the cytoskeleton. In: Modern Cell Biology, vol. 2. Spatial Organisation of Eukaryotic Cells, ed. J.R. McIntosh, pp. 115-42. New York: Alan R. Liss.

McKeithan, T.W., Lefebvre, P.A., Silflow, C.D. & Rosembaum, J.L. (1983). Multiple forms of tubulin in Polytomella and Chlamydomonas: evidence for a precursor of flagellar alpha-tubulin. Journal of Cell Biology, 96, 1056-63.

Metcalf, T.N., Szabo, L.J., Schubert, K.R. & Wang, J.L. (1980). Immunochemical identification of actin-like protein from soybean seedlings. Nature, 285, 171-2.

Morejohn, L.C. & Fosket, D.E. (1982). Higher plant tubulin identified by self-assembly into microtubules in vitro. Nature, 297, 426-8.

Nagao, R.T., Shah, D.M., Eckenrode, V.K. & Meagher, R.B. (1981). Multigene family of actin-related sequences isolated from a soybean genomic library. DNA, 2, 1-9.

Oliver, J.M., Krawiec, J.A. & Berlin, R.D. (1978). A carbamate herbicide causes microtubule and microfilament disruption and nuclear fragmentation in fibroblasts. Experimental Cell Research, 116, 229-37.

Palevitz, B.A. (1981). Microtubules and possible microtubule nucleation centers in the cortex of stomatal cells as visualised by high voltage electron microscopy. Protoplasma, 107, 115-25.

Palevitz, B.A. (1982). The stomatal complex as a model of cytoskeletal participation in cell differentiation. In: The Cytoskeleton in Plant Growth and Development, ed. C.W. Lloyd, pp.345-76. London: Academic Press.

Palevitz, B.A. & Hepler, P.K. (1974). The control of the plane of division during stomatal differentiation in Allium I. Spindle reorientation. Chromosoma (Berlin), 46, 297-326.

Persecreta, T.C., Carley, W.W., Webb, W.W. & Parthasarathy, M.V. (1982). F-actin in conifer roots. Proceedings of the National Academy of Sciences, USA, 79, 2898-901.

Pickett-Heaps, J.D. (1969). The evolution of the mitotic apparatus: an attempt at comparative ultrastructural cytology in dividing plant cells. Cytobios, 3, 257-80.

Pickett-Heaps, J.D. (1974). Plant Microtubules. In: Dynamic Aspects of Plant Ultrastructure, ed. A.W. Robards, pp. 219-55. London: McGraw-Hill.

Pickett-Heaps, J.D. (1978). Green Algae: Structure Reproduction and Evolution in Selected Genera. Massachusetts: Sinauer Associates.

Pickett-Heaps, J.D. & Northcote, D.H. (1966). Organisation of microtubules and endoplasmic reticulum during mitosis and cytokinesis in wheat meristems. Journal of Cell Science, 1, 109-20.

Powell, A.J., Peace, G.W., Slabas, A.R. & Lloyd, C.W. (1982). The detergent-resistant cytoskeleton of higher plant protoplasts contains nucleus-associated fibrillar bundles in addition to microtubules. Journal of Cell Science, 56, 319-35.

Preston, R.D. (1974). Plant cell walls. In: Dynamic Aspects of Plant Ultrastructure, ed. A.W. Robards, pp. 256-309. London: McGraw-Hill.

Robinson, D.G. & Quader, H. (1982). The microtubule-microfibril syndrome. In: The Cytoskeleton in Plant Growth and Development, ed. C.W. Lloyd, pp. 109-26. London: Academic Press.

Roobol, A., Havercroft, J.C. & Gull, K. (1982). Microtubule nucleation by the isolated microtubule organising centre of Physarum polycephalum myxamoebae. Journal of Cell Science, 55, 365-81.

Schmit, A.-C., Vantard, M., de Mey, J. & Lambert, A.-M. (1983). Aster-like microtubule centres established spindle polarity during interphase-mitosis transition in higher plant cells. Plant Cell Reports, 2, 285-8.

Simmonds, D., Setterfield, G. & Brown, D.L. (1983). Organisation of microtubules in dividing and elongating cells of Vicia hajastana Grossh. in suspension culture. European Journal of Cell Biology, 32, 59-66.

Van der Valk, P., Rennie, P.J., Connolly, J.A. & Fowke, L.C. (1980). Distribution of cortical microtubules in tobacco protoplasts. An immunofluorescence microscopic and ultra-structure study. Protoplasma, 105, 27-43.

Vahey, M., Titus, M., Trautwein, R. & Scordilis, S. (1982). Tomato actin and myosin: contractile proteins from a higher land plant. Cell Motility, 2, 131-47.

Venverloo, C.J., Hovenkamp, P.H., Weeda, A.J. & Libbenga, K.R. (1980). Cell division in Nautilocalyx explants. I. Phagmosome, pre-prophase band and plane of division. Zeitschrift für Pflanzenphysiologie, 100, 161-74.

Wick, S.M. & Duniec, J. (1983). Immunofluorescence microscopy of tubulin and microtubule arrays in plant cells. I. Pre-prophase band development and concomitant appearance of nuclear envelope-associated tubulin. Journal of Cell Biology, 97, 235-43.

Wick, S.M. & Hepler, P.K. (1980). Localisation of Ca^{++}-containing antimonate precipitates during mitosis. Journal of Cell Biology, 86, 500-13.

Wick, S.M., Seagull, R.W., Osborn, M., Weber, K. & Gunning, B.E.S. (1981). Immunofluorescence microscopy of organised microtubule arrays in structurally stabilised meristematic plant cells. Journal of Cell Biology, 89, 685-90.

Williamson, R.E. (1980). Actin in motile and other processes in plant cells. Canadian Journal of Botany, 58, 766-72.

Wilson, H.J., Israel, H.W. & Steward, F.C. (1974). Morphogenesis and the fine structure of cultured carrot cells. Journal of Cell Science, 15, 57-73.

Wolniak, S.M., Hepler, P.K. & Jackson, W.T. (1983). Ionic changes in the mitotic apparatus at the metaphase/anaphase transition. Journal of Cell Biology, 96, 598-605.

Wulf, E., Deboben, A., Bautz, F.A., Faulstich, H. & Wieland, T. (1979). Fluorescent phallotoxin, a tool for the visualisation of cellular actin. Proceedings of the National Academy of Science, USA, 76, 4498-502.

Zavortink, M., Welsh, M.J. & McIntosh, J.R. (1983). The distribution of calmodulin in living mitotic cells. Experimental Cell Research, 149, 375-85.

GROWTH SUBSTANCES, CALCIUM AND THE REGULATION OF CELL DIVISION

A.J. Trewavas

INTRODUCTION

The production of new cells is a basic component of growth and development. In higher plants most cell division is limited to meristems. These tissues, which are unique to higher plants, possess novel properties of behaviour and replication which are aptly described as plastic, and they contribute fundamentally to the overall framework of developmental plasticity found only in plants (Trewavas, 1981). Meristems represent a maintenance of the embryonic character in the adult plant; new cells are continually produced which then subsequently undergo differentiation.

In this chapter three topics are discussed which are directly relevant to the understanding of cell division and differentiation in higher plants. They are (i) the differentiation of dividing cells, (ii) growth substances and cell division and (iii) calcium-protein kinase and the cell cycle.

DIFFERENTIATION OF DIVIDING CELLS

Although dividing cells in meristems may look anatomically identical to each other and although they are obviously all undergoing at least a metabolic programme of cell replication, there is evidence to suggest that even adjacent, dividing cells may be physiologically very different to each other.

Different meristems exhibit different physiological properties

The main root and shoot meristems must be among the most stable of all plant tissues (Wareing, 1979). Despite the remarkable regenerative properties of certain cultured cells and of various plant parts, there appears at present to be no well-authenticated reports of conversion of shoot or root meristem to anything but themselves. If the shoot meristem is sub-divided by vertical cuts each regenerates to a new shoot meristem (Steeves &

Sussex, 1972); similar behaviour is seen in the root meristem.

In contrast to these, the cambium is frequently used to produce callus from which meristems can be regenerated. Simply removing the layers of cells from outside the shoot cambium can lead to unrestrained and unorganised cell division. From this it can be deduced that physical constraint is an important element in cambial organisation but not in the other shoot meristems. Other determinate organs such as leaves also have dividing cells which may easily be cultured into callus but do not show the same pressure constraints. Cortical cells may be re-initiated into division and adventitious roots regenerated. This regeneration cannot be achieved with the shoot meristem itself but only with sub-apical tissue. Wareing (1979) quotes a number of other examples of stable meristem properties which are not altered or diluted by cell division. These include vernalisation, phase changes (as in ivy) and the gametophyte/sporophyte generations in lower plants. These meristem differences must reflect physiological differences in the dividing cells of which they are composed.

Evidence that dividing cells can be differentiated

Although most cell cultures require exogenous auxin and cytokinin for growth they may, by appropriate manipulation, be switched into a mode in which they are auxin- and cytokinin-independent for cell division. Such cells are said to be habituated. Habituation can survive countless cell replications and obviously such habituated cells are physiologically different from those which are auxin- and/or cytokinin-dependent (Meins & Binns, 1978).

Perhaps the best evidence that supports the idea of differentiation of dividing cells is that of Coe & Neuffer (1978), Steffensen (1968) and their collaborators. Using anthocyanin markers in maize plants and irradiation of dry maize seeds they have been able to construct remarkable maps showing the differentiation of dividing cells in the shoot apex. Two to four cells in the shoot meristem for example are solely responsible for subsequent tassel formation; sixteen cells are responsible for the four to seven upper nodes and so on. There is great similarity between the anatomical construction of the apex and the various number of cells they have mapped as being responsible for various tissues. During germination these 'mother' cells replicate, producing clones of themselves which then subsequently form the various tissues. These 'clonal' theories of development are similar in many respects to the old histogen theories of root development. Longitudinal sections of root are strongly suggestive of lineage patterns of cells. Thus the true root meristem may represent a small population of mother cells all of which are physiologically different from each other and which clone themselves to form the major root cell types.

The interpretation of the origin of chimeras has always had a requirement that the various layers of cells in the shoot meristem are responsible for the formation of different tissues in the leaf (Stewart, 1978). Clearly these layers are physiologically different from each other and remain permanently separate in function otherwise chimeral plants could not be maintained (Stewart, 1978). Knox and Considine (1982) reported that only one to two cells are responsible for the subsequent formation of the whole epidermis in wheat florets when the organ is first initiated. These 'mother' epidermal cells must clearly be different to other cells in the floret initial.

Alongside these should be considered the relatively large number of cases of divisions which give rise to very unequally-sized daughters. Such divisions often occur at crucial developmental times and the fate of the two cells is usually very different (Trewavas 1982a). For example in the *Fucus* zygote the first division gives rise to a smaller cell which eventually generates the rhizoid and a larger cell which generates the thallus (Callow, Evans & Callow, 1982). Unequal cell divisions also occur during higher plant embryogenesis (the first division gives a smaller embryo cell and a larger suspensor cell), during pollen development, during guard cell differentiation, during phloem formation (sieve tube and companion cell), during root hair formation and during the germination of many spores or zygotes (Gurdon, 1974).

How can dividing cells be differentiated from each other?

In trying to understand how dividing cells can be different from or the same as each other there is one basic difficulty. During the cell cycle the whole of the genome is replicated. This process obviously requires an unravelling of the normal chromatin structure and a separation of DNA strands to permit replication to occur. However, this is similar to the normal processes which occur on a local scale to allow one gene to become active while others remain as part of the condensed chromatin and thus remain inactive. If the daughter cells are effectively identical then the same sets of genes must remain active/inactive after replication. How does any form of stable differentiation survive replication; how can differentiated cells retain their differences through replication? The answer may lie partly in the controlling influence of the cytoplasm.

Data from microinjection experiments suggest that in animal cells the nucleus adopts the posture of the cytoplasm in which it finds itself (Gurdon, 1974). Thus since the two daughter nuclei are exposed to the same cytoplasmic controlling elements, their regions of diffuse and dense chromatin (active vs inactive genes) must rapidly become identical. If that is the case, how then can dividing cells become differentiated from each other? The answer has come from studies on unequal divisions. By careful manipulation the two daughter nuclei of an unequal division have been rotated during telophase and the fate of the two subsequent unequal daughter cells followed (Gurdon, 1974). No differences were observed from the normal unmanipulated condition; thus clearly at cytokinesis the daughter nuclei are identical. The subsequent different fates of the two cells must then be determined by fundamental differences in the cytoplasms the two nuclei find themselves in.

Further evidence for unequal partitioning of gene-regulating elements at mitosis comes from studies of plant meristems. In the root cap initials of *Zea mays*, the cells and their nuclei are larger in Row I than the corresponding sister cells and nuclei in Row II (Davidson, Pertens & Eastman, 1978; Ivanov, 1979). In roots treated with caffeine, an inhibitor of cell plate formation (Gimenez-Martin, Gonzalez-Fernandez & Lopez-Saez, 1965), the pairs of sister nuclei in the resulting binucleate cells show the same differences in size and polarity, as control sister cells *i.e.* in the binucleate cell the larger nucleus of a sister pair lies in what would be the Row I position had the cell been allowed to divide normally (Davidson & Pertens, 1978). Furthermore, during the subsequent interphase, the pairs of sister nuclei show differential growth with the actual difference in nuclear size increasing with time (Davidson & Pertens, 1978). The authors conclude that firstly, a gradient existed in the mitotic cell for some factors that control nuclear volume, and secondly the factors are polarised so that the higher concentration is at the pole of the cell which will give rise to the Row I cell. Thirdly, these factors are either complexed with the anaphase chromatids or they become incorporated into the newly-formed nucleus at the end of telophase, and fourthly, once formed the nuclei retain these factors and subsequent nuclear behaviour is autonomous even when nuclei occupy a common cytoplasm. The nucleus sheds its RNA and non-histone proteins during mitosis and these macromolecules move back into the daughter nuclei near, or at the completion, of division (Phillips, 1972; Goldstein, 1976; Rao & Prescott, 1970). It therefore appears unlikely that these factors controlling nuclear volume, were complexed to the chromatids, but were more likely to have been unequally distributed in the cytoplasm at the time of mitosis and move back into the daughter nuclei as they reformed at the end of telophase. In these cases then the cytoplasm must be polarised, the gene-regulating elements are unequally partitioned and must be non-diffusable. We know from studies on Fucus that a major factor determining this unequal partitioning is the development of an electrical potential between the two ends of the cell which polarises the charged cytoplasmic and membrane bound protein components (Jaffe, 1969; Trewavas, 1982a). We also know that asymmetric cytoplasmic calcium distributions are responsible for the development of the electrical potential. Calcium is thus a crucial element in establishing the differentiation of dividing cells.

CELL DIVISION, REGENERATION AND GROWTH SUBSTANCES

The notion that auxin and cytokinin act as key regulators of cell division in the intact plant is firmly entrenched. More specifically it is assumed that movements of auxin and cytokinin from elsewhere in the plant act to determine the level and extent of cell division in meristems. The evidence for these ideas seem to be based on the recognised exogenous auxin and cytokinin requirements for division exhibited by many cultured cell lines and on the general assumption that growth substances are the only specific and limiting regulators of the division process. This latter notion seems to derive in turn as much from observations on the manipulation of root or shoot regeneration and embryogenesis by auxin and cytokinin as anything else and, additionally, from observations on the effects of very high concentrations of auxin on cell division in intact plants.

Most cultured cells are derived from mature cells which have been re-initiated into division by various treatments. As pointed out earlier the shoot and root meristem have rarely, if ever, been cultured into callus and it is not known whether the division in these areas is growth substance-dependent in the same way. The technique of culture, particularly of cells and small tissue pieces, has hazards in interpretation because the culture medium may provide a considerable diffusion gradient for endogenous growth substances which is not present in the intact plant. Cultured cells do synthesise their own auxin and cytokinin (Sembdner, Gross, Webisca & Schneider, 1980) and larger pieces of tissue, e.g. root, may be cultured and grown without a requirement for exogenous growth substances. Thus meristems in the intact plant may not be dependent on transported growth substances to maintain cytokinesis.

Plant cells can be maintained in a state of division without exogenous growth substances. Crown gall cells and habituated cultures are two examples. However, until recently, it has always been assumed that these cells synthesise their own auxin and cytokinin in much greater quantities than non-dividing tissue to maintain cell division. Weiler & Spanier (1981) analysed over 30 crown gall cultures and found that, providing the culture was sterile, the level of these two growth substances in many of the cultures was no higher than the non-dividing parent tissue from which they were derived. Furthermore, Nakajima _et al._ (1979) found the levels of auxin and cytokinin in three habituated cultures to be almost undetectable. If auxin and cytokinin are essential for cell division then these results suggest that sensitivity to auxin and cytokinin is altered in crown gall and habituated cells. Thus, it is the change in sensitivity and not a change in the level of growth regulators which is important.

Many factors regulate cell division and regeneration

Table 1 summarises the various factors which have been found to induce cell division in various tissues (bean pod for example), to specifically modify regeneration in callus or epidermal explants, to promote cell division leading to adventitious root formation in shoots or to modify membrane potential. I have set out this information to counter the notion that growth substances are the only endogenous materials capable of specifically modifying development and cell division. In this connection the following observations taken from Tran Thanh Van (1981), Tran Thanh Van, Chlyah & Chlyah (1974), Tran Thanh Van & Trinh (1978) and Kohlenbach (1978) seem of relevance. In cultured cells in a medium containing auxin and cytokinin, low concentrations of ammonia are required for embryo formation; if the medium only contains nitrate, only callus develops. Changing from glucose to sucrose in the culture medium initiates tracheid formation in artichoke or regenerates floral buds instead of vegetative buds. Increasing the osmotic pressure of the culture medium with mannitol greatly increases the number of shoots regenerating. Potassium chloride at 1 mM in the medium initiates callus formation; 20 mM initiates embryo formation. Increasing the temperature at which the parent plant is maintained from 17° to 24° leads to massive increases in subsequent hair formation by daughter explants. Substituting

TABLE 1

A comparison of the factors inducing cell division in bean pod and other tissues, specifically modifying regeneration from callus or epidermal strips, specifically promoting adventitious root primordia formation or modifying membrane potential

Callus or epidermal layer regeneration	Roof formation	Cell division in bean pod or shoot bud	Membrane potential
osmotic pressure	water (osmotic pressure)	osmotic pressure	osmotic pressure
various minerals (K^+, NH_4^+ NO_3^-)	various minerals	various minerals	various minerals
sucrose	sucrose	sucrose	sucrose
vitamins	vitamins	-	-
auxin, cytokinin, gibberellin ethylene	auxin, cytokinin, ethylene, abscisic acid	auxin, cytokinin, ethylene	auxin
various organic substances, polyamines amino acids	various organic compounds amino acids	various organic compounds polyamines amino acids	fusicoccin
-	uncouplers, fatty acids dinitrophenol	fatty acids, dinitrophenol	dinitrophenol
peptides, proteins	-	peptides, proteins	-
phenols	phenols	-	-
light	light	light	light
pH, CO_2	pH	pH, CO_2	pH
temperature	temperature	temperature shock	temperature
mechanical effects contraint	wounding	wounding or pressure	mechanical force

References: Kahl (1973), Fernqvist (1967), Tran Thanh Van (1981), Trewavas (1982), Hess (1968), Street (1968).

The above should be compared with the factors inducing parthenogenesis in sea urchin eggs, specifying polarity in *Fucus*, breaking seed dormancy, initiating bacterial tumbling: see Trewavas (1982a,b,c, 1983a). The similarity in all these lists suggests a common site of action in all these cases of development. As pointed out earlier this is probably the permeability of membranes to calcium ions (Trewavas,1982a).

indolebutyric acid with 2,4-dichlorophenoxyacetic acid prevents root regeneration. Whereas auxin and cytokinin induce cell division in artichoke cells in the presence of mineral salts, in the absence of minerals they only induce cell expansion; minerals in this case are the critical cytokinetic factor. With tobacco genotypes difficult to regenerate, similar difficulty is found with regeneration in cultures; culture conditions thereby only permit expression of a predetermined potential for regeneration. In tobacco, a high ratio of auxin to cytokinin encourages root regeneration and a low one, shoot regeneration; in Medicago it is the complete reverse.

Since the above manipulations are carried out in the presence of exogenous growth substances there is no reason to suppose that these conditions specifically modify the biosynthesis of auxin or cytokinin which then in turn lead to specific developmental changes. Obviously a variety of factors are essential for cell division or regeneration. Any of these can be made limiting and thus in culture (and probably in the whole plant) used to control development. It makes no sense to pick on one and elevate that to the level of ultimate controller.

Plants live in a constantly varying environment and inevitably many of their internal constituents, for example minerals, photosynthate and water, vary with it. As Table 1 shows many of these affect cell division. Is it practicable to assume that growth substances are always limiting factors in cell division or are they ever limiting? Day-to-day variation in other required materials (sugar, minerals, water) will constantly change the responsiveness of dividing cells to growth substances. Could a system controlled only by one substance ever evolve when the scale of the response could never be guaranteed?

These questions are very relevant to the control of the cambium. For many years it was assumed that the concentration of auxin was the main regulator of cambial cell division activity. These notions were derived from experiments in which applications of very high concentrations of auxin induced apparent cambium formation and from the supposed effects of bud-produced auxin on cambial activity (Sinnott, 1960). However cambium can synthesise its own auxin (Sheldrake, 1973). In addition, the apparent congruity between the onset of bud activity in springtime and the initiation of cambial cell divisions below the tree bud (supposedly auxin moving in a polar direction) is countered by the observations that for many trees, springtime cambial activity starts from the base of the tree and moves upwards or in others goes down the main shoot and up into the new shoots (Sinnott, 1960;

Perry, 1971). Other trees apparently do not need buds for cambial division activity at all. Many factors, including water availability, various minerals, photosynthate and light conditions, growth substances and time of year are known to modify division activity in the cambium (Perry, 1971). Compare these with Table 1. The most reasonable conclusion to draw from these observations is that the movement of a variety of materials through the adjacent vascular tissue is the stimulus renewing and controlling cambial activity. This conclusion however ignores the observations that suggest autonomy in cambial activity and an obvious propensity in cambial cells to commence cell division and growth in springtime. This will now be considered.

VARIATIONS IN SENSITIVITY TO GROWTH SUBSTANCES CONTROLLING CELL DIVISION AND REGENERATION

One of the major problems with experiments manipulating cell division or regeneration in culture by growth substances are the conceptual implications they suggest. The cell is seen as something like a lump of clay moulded by various levels of growth substances into different types or forms. Such a view ignores any input into the process of development by the cell itself and yet consideration of a few examples suggests that such an input may be paramount.

1. As very young coleoptile cells prepare to enter a phase of cell division they acquire sensitivity to cytokinins. The variations in sensitivity slightly precede and accurately describe changes in the number of dividing cells (Trewavas, 1982b).
2. When artichoke tuber discs are first excised they respond to auxin and cytokinin application by entering cell division. If the discs are washed for 24 hours first then they respond to added auxin and cytokinin only by expansion (Setterfield, 1963).
3. Cambial cells which undergo an annual cycle in division activity acquire sensitivity to auxin in springtime just prior to the onset of division (Gouwentak, 1941). These variations can occur in the apparent absence of environmental change.
4. Cells show a continual variation in their sensitivity to growth substances as they develop and mature. If etiolated soybean hypocotyls are treated with auxin the meristem ceases division, the elongating zone ceases elongation whilst the cells which have just matured recommence division and eventually produce adventitious root primorda (Key & Shannon, 1964). Fernqvist (1967) examined serial sections of mungbean and found a steep gradient in the

initiation of cell division induced by auxin and subsequent root primordia formation. The numbers of such primordia reach a peak about 3 cm below the cotyledons and then decline. Tran Thanh Van and Trinh (1978) examined serial epidermal strips in tobacco and found a continuous gradient in growth substance-induced regeneration. Tran Thanh Van (1981) and Dore (1965) list a massive number of examples of age, time and tissue related sensitivity variations in growth substance-induced regeneration.

5. Dormant buds in potato tubers undergo a continuous variation in sensitivity to cytokinins (C. Turnbull & D. Hanke, personal communication). Cytokinin sensitivity increases until dormancy is broken and cell division commences.

6. Early somatic embryogenesis in soybean is promoted by exogenous abscisic acid. As embryos continue development, however, abscisic acid becomes inhibitory (Ackerson, 1984).

7. Most of the examples quoted in the previous section result from sensitivity changes. Thus for example in terms of floral bud regeneration, sucrose drastically alters the sensitivity of epidermal explants to auxin and cytokinin (Tran Thanh Van, 1981).

Although these observations are limited by the requirement to consider only cell division they are part of a general pattern which is seen in plant development (Trewavas, 1981, 1982b). That is, the acquisition of sensitivity (or competence) precedes and controls the act of development itself; it determines how and when and the degree to which the tissue can respond.

Developmental processes can be divided into two broad phases, specification (competence, sensitivity) and expression (realisation) (Mohr, 1982). What the above examples show is that cells are specified (acquire sensitivity) for an act of development before its expression. As specification (sensitivity) is acquired, if growth substances and other environmental factors are present, expression of that specification occurs concomitantly; a situation which frequently happens in the intact plant. Are growth substances anything to do with specification? With only five available it would be simplistic to suppose that they were the only major sources of information necessary to specify the development of even the simplest tissue. As cells are specified to become dividing cells part of that programme of specification is the acquisition of sensitivity to cytokinin or auxin. If auxin and cytokinin are freely available in the medium what is then seen is the appearance of dividing cells.

Because specification has not usually been distinguished from expression, auxin and cytokinin are often talked of as inducing cell division. It is more accurate to say that auxin and cytokinin are essential factors for previously-specified cells to accomplish division at a reasonable rate. As Table 1 shows that many factors in addition to growth substances are needed for expression. It is however the acquisition of sensitivity, the specification, which is the limiting factor determining whether the cells are able to divide at all. Quite simply, then, a factor which may be essential for the specification to be expressed is being confused with the control of specification itself. In regeneration from epidermal explants, the dividing cells are evidently differentiated, each perhaps from different primordial mother cells. The particular balance of materials in the medium subsequently ensures the expression of one of those specifications giving rise, for example, to roots or buds.

Do growth substances regulate division in the intact plant?

What are growth substances for? From other studies it is evident that growth substances are coupled to the expression of a group of genes which, when expressed, circumvent previously-limiting steps in metabolism (Trewavas, 1976b, 1982a). Although in evolutionary terms they may have been used to speed up the division process they may now be essential in some plants for the cell cycle to occur at all. However it is development which specifies where and usually when division will occur. If growth substances are to solely regulate the expression of that specification they must consistently be the limiting factor in the whole division process. Faced with so many factors for division (Table 1) this is frankly difficult to believe. In the intact growing plant the limits on cell division and thus subsequent growth seem to be mainly water, temperature, minerals and light. Growth substances are thus part of the genetic potential for rapid cell division but the extent to which that potential is realised depends upon the availability of other resources. Under optimal growing conditions added growth substances have little or no effect on increasing the rates of cell division.

In other situations dealing with effective initiation of cell division (e.g. apical dominance, rosette plants, branch root formation) the contribution of sensitivity to growth substances seems paramount (Trewavas, 1981, 1982b). All measurements of growth substance concentrations have failed to establish that endogenous levels are in any way limiting and that initiation is accomplished by increasing the endogenous level. The evidence which has previously been used to support the notion that growth substances are limiting has relied heavily on results obtained with grossly excessive concentrations of applied growth substances. These applied levels are rarely less than four and often more than six orders of magnitude higher than endogenous levels of growth substances. What do these data mean?

A cell is a complex, interlocking, network of dependent and interdependent metabolic systems regulating gene expression. Disruption, dislocation or mismatching of some of these, as inevitably follows severe experimental treatments, is difficult if not impossible to interpret as regards the endogenous control. Again many tissues in a plant interact. The possibility that a specific response occurs because of effects of exogenous growth substance on the metabolism of other tissues seems never to be considered.

Growth substances may have some parallels with cyclic AMP in bacteria. Cyclic AMP is accumulated internally and used to co-ordinate the expression of a disparate number of genes which initiate a metabolic behavioural response of the bacterial cell to particular growth conditions. However the response in this case is controlled by varying the level of a small molecule because it must be completely reversible; the bacterial cell can completely return to its initial state when environmental conditions improve. The same is true for mammalian hormones. However from the first division of a plant cell in a meristem, a continual programme of developmental change is undergone, generally through declining division rates, increasing and decreasing expansion and then variation in maturation as well as other specific tissue differentiation. This involves a continuous change in sensitivity to growth substances and except under cultures conditions is not reversible. These observations lead naturally to the conclusion that sensitivity to growth substances is one of the specific and controlling factors in development.

CALCIUM, PROTEIN KINASES AND CELL DIVISION

The calcium hypothesis

The calcium hypothesis is summarised in Figure 1. The evidence supporting this hypothesis comes from direct measurements of cyto-plasmic calcium, introduction of Ca^{2+} into cells by ionophoresis or microinjection and calcium requirements for growth of various oncogenic cell cultures in different media, (Berridge, 1976; Roux & Slocum, 1982; Dunham & Walton, 1982). As a general rule, in cells which are pre-programmed for cell division, the cell cycle is induced by an increase in cytoplasmic calcium to the 1-10 μM range. In plants there are a variety of ways of doing this. These include various shock treatments or treatment with various (usually hydrophobic) chemicals including growth substances (Trewavas, 1982a). The effect of these treatments on the calcium permeability may be mediated by the voltage across the plasma membrane. Calcium enters the cell from outside via specific proteins (so-called calcium channels) and the functioning of these is controlled directly by the membrane voltage and phosphorylation (see references in Hetherington & Trewavas, 1984a). Anything affecting proton motive force (e.g. dinitrophenol, uncoupler) is likely to affect calcium permeability.

How do elevated cytoplasmic calcium levels initiate the cell cycle?

Cytoplasmic calcium levels are sensed by a variety of proteins which bind calcium to differing extents. The most well-known of these is calmodulin. Calcium is bound by calmodulin at concentrations of 10^{-7} to 10^{-6}M and its conformation is altered. In this form it can bind with and activate numerous enzymes. In plants four of these have been identified: NAD kinase (Anderson & Cormier, 1978), Ca^{2+} ATPase (Dieter & Marme, 1980), quinate oxidoreductase (Ranjeva, Refeno, Boudet & Marme, 1983) and protein kinase (Hetherington & Trewavas, 1982, 1984b). Out of these, protein kinase is probably the most important because of its pleiotropic function. Both soluble and membrane-bound, calcium-activated protein kinases have been detected (Polya & Davies, 1982; Hetherington & Trewavas, 1982).

Protein kinases appear to regulate steps in the cell cycle of plants and animals

Recent studies have suggested that changes in protein kinase

Fig. 1. The calcium hypothesis.

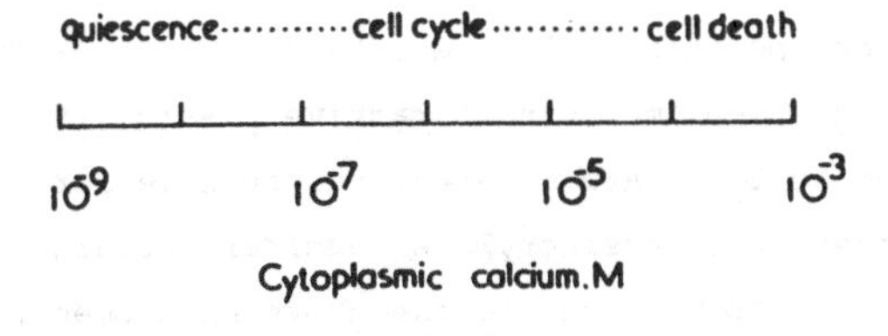

activity and the pattern of phosphorylated nuclear proteins accompany the process of cell division. Thus for plants:

A. Nuclei isolated from the mitotic zone of pea roots have a three-fold higher level of protein kinase activity than nuclei from the mature non-dividing zone. In addition they possess four additional major phosphorylated nuclear proteins not detectable in nuclei from non-dividing cells (Trewavas, 1979a).

B. A prominent and rapid feature of treatment of *Lemna* with abscisic acid is the inhibition of cell division. Untreated and therefore dividing controls contain two-fold higher nuclear protein kinase activity and at least one additional major phosphorylated nuclear protein (Chapman, Trewavas & Van Loon, 1975).

C. In barley embryo cell nuclei, the approach of mitosis (late G2/M) is accompanied by the transient appearance of six phosphorylated nuclear proteins (Trewavas, 1979b).

D. Histone H1 in cultured artichoke tuber cells shows a marked increase in covalently-attached phosphate upon the approach of mitosis. In *Physarum* it has been shown that this increased phosphorylation is responsible for chromatin condensation and is caused by a 20-fold increase in histone H1 kinase activity (Bradbury, Inglis & Matthews, 1974; Trewavas, 1976a; Stratton & Trewavas, 1981).

E. Artichoke tuber cells induced to divide by excision and auxin treatment increase their nuclear protein kinase activity about two-fold just prior to S-phase. The initiation of DNA synthesis is accompanied by the appearance of four to six novel phosphorylated nuclear proteins (Melanson & Trewavas, 1981; Costa, Fuller, Russel & Gerner, 1977).

Since several protein kinases are present in nuclei, the two- to three-fold changes in activity may disguise much larger changes in just one or two. Although the functions of the nuclear phosphorylated proteins which change are not known, they usually represent major nuclear species and a reasonable hypothesis would suppose them to be enzymes or structural proteins concerned with the specific nuclear events of the cell cycle phase (see Dunham & Bryant, this volume). An attractive possibility is that the changes in protein kinase activity are directly responsible for the specific alterations in the patterns of phosphorylated nuclear proteins.

Similar observations to the above have been made with animal cells. Both the pattern of nuclear protein kinase activity and phosphorylated nuclear proteins change through the cell cycle (Costa *et al.*, 1977; Philips _et al._, 1979). More significantly the direct regulatory role of protein kinase in the cell cycle has been clearly demonstrated. Addition of exogenous protein kinase to cultured cells greatly increases the proportion initiating DNA synthesis (Boynton & Whitfield, 1980, and refs. therein). Some of the enzymes concerned with DNA replication have been shown to be phosphorylated (DNA polymerase for example) and to appear in the nucleus during S-phase (Trewavas, 1976a; Moreau, Guerrier & Doree, 1976; Jordan, Timmis & Trewavas, 1980). Treatment of _Physarum_ with exogenous histone H1 kinase can actually accelerate the onset of mitosis and similar results have been obtained by microinjection of protein kinase into oocytes. Calmodulin-regulated protein kinases have also been detected in nuclei, (Sikorska, MacManus, Walker & Whitfield, 1980; Landt & McDonald, 1980) and this would agree with the data discussed later which suggest calcium to be a specific regulator of certain cell cycle steps.

Tyrosine kinase and the cell cycle

Until recently it was thought that the only amino acids phosphorylated in proteins were serine, threonine and possibly lysine and histidine. However a new set of protein kinases phosphorylating tyrosine have recently been discovered and although the level of tyrosine phosphorylation is often low these enzymes have been found to be critical to the process of the cell cycle.

Epidermal growth factor, insulin and platelet-derived growth factors are all peptide hormones found in mammals. Added to appropriate cell cultures they cause increases in respiration rate and increases in RNA and protein synthesis leading to replication of DNA and cell division (Hetherington & Trewavas, 1984a).

Such changes in metabolism also occur in storage tissue discs incubated in water, in lateral buds freed from apical dominance, in cortical cells induced to divide by added growth substances, in dormant buds reactivated by ethylene chlorhydrin and in embryo germination (Trewavas, 1982a).

Addition of epidermal growth factor or insulin to isolated membrane preparations results in three-to five-fold increases in associated protein kinase activity. The receptors for all three hormones have been purified and have been shown to be tyrosine kinases. The main substrate for increased phosphorylation is the receptor, tyrosine kinase itself, which thus autophosphorylates. This has been clearly shown by the use of the affinity label, azido-ATP to demonstrate the identity of receptor with enzyme function.

After phosphorylation the increased tyrosine kinase activity is independent of further combination with the hormone.

Other evidence supporting the critical role of tyrosine kinase in the cell cycle has come from a totally different direction (Hetherington & Trewavas, 1984a). The transforming viruses are a set of RNA-containing viruses which are able to transform ordinary cultured mammalian cells with full cell cycle control into cancerous cells in which cell replication is permanent. The Rous sarcoma virus (src) is the most well known of these. The transforming ability of these viruses has been located to one gene, the so-called oncogene. In the Rous sarcoma virus this gene codes for a product of 60,000 molecular weight which it itself phosphorylated and is a tyrosine kinase loosely attached to the plasma membrane. After transformation, tyrosine kinase levels are permanently elevated some five-to ten-fold and this is sufficient to switch the cell into continuous cell replication.

The oncogene products of at least eight other retroviruses are tyrosine kinases which group into four families. There is evidence for movement of tyrosine kinase between the membrane and the soluble fraction. To date only one substrate for the tyrosine kinase has been found and that is vinculin, a protein concerned with the attachment of microfilaments to the cell membrane.

These separate lines of evidence argue strongly for a fundamental role of tyrosine kinase in the cell replication process in animals. This role may be that of controlling calcium entry. Calcium entry takes place through membrane-located protein pores (calcium channels) whose functioning is voltage regulated and controlled by phosphorylation. Direct evidence shows epidermal growth factor to induce calcium entry into recipient cells and much evidence associates oncogenesis with a permanent elevation of cytoplasmic calcium levels (Dunham & Walton, 1982). A simple hyopthesis suggests therefore that one of the substrates for tyrosine kinase could be the calcium channel itself. Microinjection of protein kinase has been shown to modify calcium influx (Osterreider *et al*., 1982). These notions are summarised in Figure 2. The possible presence of tyrosine kinase in plants is being actively pursued and preliminary results suggest it is present (D. Blowers & A.J. Trewavas, unpublished data).

Protein kinase-C, the PI effect, Ca^{2+} and cell replication

A large number of stimulating factors (e.g. epidermal growth factors, insulin etc.) which lead to cell replication in animal cells have been found to induce the turnover of phosphatidyl inositol in the membranes of the treated cells. This effect is known as the PI effect and it is believed to result from the increased activity of phospholipase-C in the membrane. Michell (1983) has argued that the PI effect may be responsible for increased Ca^{2+} entry leading to cell replication. However there are numerous instances where increased Ca^{2+} entry appears to precede the PI effect.

The action of phospholipase-C on phosphatidyl-inositol is to generate diglycerides. Recently a protein kinase has been detected and purified which is specifically activated by diglyceride and by calcium ions. In the absence of diglyceride the optimal Ca^{2+} concentration is 10^{-4} to 10^{-3}M but in the presence of diglyceride this drops to 10^{-6} to 10^{-5}M. This protein kinase-C is an obvious candidate for the transduction of elevated levels of diglyceride and Ca^{2+}, leading to cell activation.

Phorbol esters are chemicals isolated from members of the Euphorbiaceae and which act as tumour promoters; they are not oncogenic themselves but they greatly potentiate the effects of oncogenic chemicals. The phorbol ester receptor has been purified and is in fact protein kinase-C (Michell, 1983). Addition of phorbol esters to homogenates or to whole

Fig. 2. Summary of possible inter-relations of tyrosine kinase and calcium entry.
EGF = epidermal growth factor
TK = tyrosine kinase
Src mRNA = Rous sarcoma oncogene mRNA

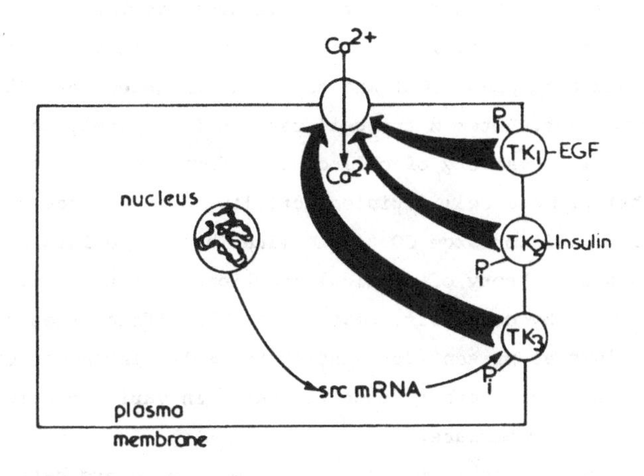

cells leads to an attachment of protein kinase-C to the plasma membrane. The substrates for the enzyme are unknown but are clearly proteins and enzymes concerned with the business of cell replication.

The PI effect has been detected in plants (D. Hanke, personal communication) and the possible presence of protein kinase-C is being actively pursued. Recently detected membrane-bound calcium-activated protein kinases do not seem to be protein kinase-C (Hetherington & Trewavas, 1984b).

CALCIUM AND THE CELL CYCLE: CONCLUDING REMARKS

Numerous concepts have been developed to try and explain the very obvious repetitious nature of the cell cycle itself. The traditional one is a sequence of reactions: perhaps, simply, the slow linear readout of a set of genes. Thus S follows G1, G2 follows S, M follows G2 and so on, each phase determined by a unique complement of gene products. However S and M and other facets of the cell cycle seem to be separable events; for example, many plant cells continue DNA replication in the absence of mitosis (see Nagl, Pohl & Radler, this volume). S-phase can be dissociated from the increase in mass and size that is usually associated with the cell cycle (Trewavas, 1982c). Pringle & Hartwell's (1981) genetic analysis of yeast cell cycle temperature-sensitive mutants suggests the operation of at least six parallel sequences. Each sequence involves the discrete expression of a set of genes which are transcribed and translated sequentially with complex inter-relations between the various sequences.

A separate model of the cell cycle is that of transition probability (Smith & Martin, 1973). It is assumed in this hypothesis that cells are in a quiescent phase (GO) and are then triggered into G1 by the operation of a random event. After M the cell reverts immediately to GO. The model can be shown to fit a variety of cell cycle information and it is assumed that factors that promote cell division activity do so by increasing the probability of transition from GO to G1. (The idea is reminiscent perhaps of the transition state theory of chemical reactions with the reduction of activation energy produced by catalysts, heat etc.). The difficulties of the concept are the failure at present to identify the molecular event of transition (Bryant, 1982), but there is possible basis in variations in ion conductances across the plasma-membrane.

Two further concepts of the cell cycle are based on oscillatory reactions. The concept of a dividing cell as an oscillating system follows naturally from the repetitious nature of the event itself. The first may be likened to a relaxation (syphon) kind of rhythm. It is assumed that cell cycle events are coupled to a critical cell size and are initiated when this size or critical mass is reached (Nurse, 1980). Many years ago Commoner & Zucker (1953) were able to show a constant relation between cell surface area and anthocyanin accumulation in Coleus cells of very different sizes. Clearly some sort of size monitor is present in plant cells. However the difficulty with this as a general theory, as Gilbert (1981) has pointed out, is that different species have replicating cells of very different sizes.

The second concept relies on a control system for cell division being a limit cycle. A limit cycle oscillation is one which is stable, under fixed conditions and independent of the starting concentrations of the components, whose levels vary rhythmically with time. The main proponent of this view is Gilbert (1977, and references therein, 1981). He points out that any model of the cell cycle should account for the many different ways in which dividing cells can be synchronised. These include for example, heat or cold shock, inhibitors, chemical treatment. It should also account for the bewildering array of chemical reagents, nutrients and regulators and other conditions and the various interactions between these on the length of the cell cycle and the number of cells undergoing division. Gilbert (1977, 1978) has argued that some sort of overall, oscillating, control process is capable of explaining these phenomena and has suggested that the controlling system is one of reducing potential. This consists of reactions in which thiols and disulphides are synthesised and degraded, oxidised and reduced. Some evidence supports an oscillation in these parameters through the cell cycle.

That a variety of factors initiate cell division in plants has already been indicated (see Table 1 and Trewavas, 1982a, 1983b). These factors have been interpreted as having effects on membrane potential, particularly with regard to calcium influx and proton motive force. Thus heat shocks which are used to synchronise cells would be expected to result in a greatly increased calcium influx. Oscillations in cytoplasmic Ca^{2+} levels are already known to occur in heart cells arising from a simple feedback system of uptake with a membrane-bound calcium-activated protein kinase (Trewavas, 1976a). There are cytoplasmic rhythms in the cell cycle involving contraction of the plasma membrane (Gilbert, 1981); calcium is known to be a definite controlling factor in contractile processes (Berridge, 1976). Cell cycle variations in intracellular pH have been observed reflecting perhaps the known coupling of calcium uptake with proton motive force. In addition, the two major events in the cell cycle, S-phase and mitosis, are directly associated with protein phosphorylation changes and thus protein kinase modifications which seem to be controlled by calcium.

All of this argues for oscillations in cytoplasmic calcium and proton motive force as being at least one, basic and perhaps, the controlling, rhythmic event in the cell cycle.

REFERENCES

Ackerson, R.C. (1984). Regulation of soybean embryogenesis by abscisic acid. *Journal of Experimental Botany*, 35, 403-13.

Anderson, J.M. & Cormier, M.J. (1978). Calcium dependent regulator of NAD kinase in higher plants. *Biochemical and Biophysical Research Communications*, 84, 595-602.

Berridge, M.J. (1976). Calcium, cyclic nucleotides and cell division. *Symposia of the Society for Experimental Biology*, 30, 219-33.

Bradbury, E.M., Inglis, R.J. & Matthews, H.R. (1974). Control of cell division by very lysine-rich histone (f1) phosphorylation. *Nature*, 247, 257-61.

Bryant, J.A. (1982). DNA replication and the cell cycle. In *Encyclopaedia of Plant Physiology, 14B*, eds. B. Parthier & D. Boulter, pp.75-110. Berlin: Springer-Verlag.

Boynton, A.L. & Whitfield, J.F. (1980). A possible involvement of type II cAMP dependent protein kinase in the initiation of DNA synthesis by rat liver cells. *Experimental Cell Research*, 126, 477-81.

Callow, M.E., Evans, L.V. & Callow, J.A. (1982). Fucus. In: *The Molecular Biology of Plant Development*, eds. H. Smith and D. Grierson, pp. 159-85. Oxford: Blackwell Scientific Publications.

Chapman, K., Trewavas, A.J. & Van Loon, K. (1975). Regulation of the phosphorylation of chromatin-associated proteins in *Lemna* and *Hordeum*. *Plant Physiology*, 55, 293-8.

Coe, E.H. & Neuffer, R.G. (1978). Embryo cells and their destinies in the corn plant. *Symposia of the Society for Developmental Biology*, 36, 113-29.

Commoner, B. & Zucker, M. (1953). Cellular differentiation: an experimental approach. In: *Growth and Differentiation in Plants*, ed. W.E. Loomis, pp. 339-93. Iowa: Iowa University Press.

Costa, M., Fuller, P.J.M., Russel, D.H. & Gerner, E.W. (1977). Cytoplasmic and nuclear protein kinases during the cell cycle. *Biochimica et Biophysica Acta*, 479, 416-26.

Davidson, D. & [illegible]rtens, E. (1978). Differences in volumes of sister nuclei in binucleate cells: Evidence for asymmetry of mitosis. *Canadian Journal of Botany*, 56, 2363-69.

Davidson, D., Pertens, E., & Eastman, M.A. (1978). Nuclear and cell sizes in different regions of root meristems of *Zea mays* L. *Annals of Botany*, 42, 1429-38.

Dieter, P. & Marme, D. (1980). Calmodulin activation of plant microsomal Ca^{2+} uptake. *Proceedings of the National Academy of Sciences, USA*, 77, 7311-4.

Dore, J. (1965). Regeneration. In: *Encyclopaedia of Plant Physiology (2)*, ed. A. Lang, pp. 1-92. Berlin: Springer-Verlag.

Dunham, A.C.H. & Walton, J.M. (1982). Calcium ions and the control of proliferation in normal and cancer cells. *Bioscience Reports*, 2, 15-30.

Fernqvist, I. (1967). Studies on the initiation of adventitious roots. Lantbrukshogskolaus Anneler, 32, 109-244.
Gilbert, D.A. (1977). Density dependent limitation of growth and the regulation of cell replication by changes in the triggering level of the cell cycle switch. Biosystems, 9, 215-28.
Gilbert, D.A. (1978). Feedback quenching as a means of effectively increasing the period of biochemical and biological oscillators. Biosystems, 10, 227-33.
Gilbert, D.A. (1981). Cellular oscillations: relative independence of enzyme activity rhythms and periodic variations in the amount of extractable protein. South African Journal of Science, 77, 541-6.
Gimenez-Martin, G., Gonzalez-Fernandez, A. & Lopez-Saez, J.G. (1965). A new method of labelling cells. Journal of Cell Biology, 26, 305-9.
Goldstein, L. (1976). Role for Small Nuclear RNAs in "Programming" Chromosomal Information. Nature, 261, 519-21.
Gouwentak, C.A. (1941). Cambial activity as dependent on the presence of growth hormone and non-resting conditions of stems. Proceedings of Koniklijke Nederlandse Akademie van Wetenschappen 44, 654-63.
Gurdon, J. (1974). The control of gene expression in animal development. Oxford: Clarendon Press.
Hess, C.E. (1968). Factors controlling adventitious root formation. In: Root Growth, ed. W.J. Whittington, pp. 42-54. New York: Plenum.
Hetherington, A. & Trewavas, A.J. (1982). Calcium dependent protein kinase in pea shoot membranes. FEBS Letters, 145, 67-71.
Hetherington, A. & Trewavas, A.J. (1984a). Regulation of membrane bound protein kinases by phospholipid and calcium. Phytochemical Society Symposia, (in press).
Hetherington, A. & Trewavas, A.J. (1984b). Regulation of membrane-bound protein kinase in pea shoots by calcium ions. Planta (in press).
Ivanov, V.B. (1979). Determination of Sequence of Emergence of Separate Cells from the Mitotic Cycle and their transition of Depositing Starch in the Sheath of Maize Roots. Doklady Akademii Nauk. SSSR, 245, 716-9.
Jaffe, L.F. (1969). On the centripetal course of development: the Fucus egg and self-electrophonesis. Developmental Biology, Supplement 3, 83-111.
Jordan, E.G., Timmis, J.N. & Trewavas, A.J. (1980). The Nucleus. In: The Biochemistry of Plants - a Comprehensive Treatise, Vol. 1, eds. E. Talbot, P. Stumpf & E. Conn, pp. 490-590. London: Academic Press.
Kahl, G. (1973). Genetic and metabolic regulation in differentiating plant storage tissue cells. Botanical Reviews, 39, 281-321.
Key, J.L. & Shannon, J.C. (1964). Enhancement by auxin of RNA synthesis in excised soybean hypocotyl tissue. Plant Physiology, 39, 360-4.
Knox, R.B. & Considine, A.J. (1982). Deterministic and probabilistic approaches to plant development. In: Axioms and Principles of Plant Construction, ed. R. Sattler, pp. 112-7. Hague: Nijhoff/ Junk.
Kohlenbach, H.W. (1978). Comparative somatic embryogenesis. In: Frontiers of Plant Tissue Culture 1978, ed. T.A. Thorpe, pp. 59-67. Calgary: International Association for Plant Tissue Culture.
Landt, M. & McDonald, J. (1980). Calmodulin activated protein kinase activity of adepocyte microsomes. Biochemical and Biophysical Research Communications, 93, 881-8.

Meins, F. & Binns, H.N. (1978). Epigenetic clonal variation in the requirement of plant cells for cytokinins. Symposia of the Society for Developmental Biology, 36, 187-200.

Melanson, D. & Trewavas, A.J. (1981). Changes in tissue protein pattern associated with the induction of DNA synthesis by auxin. Plant, Cell and Environment, 5, 53-64.

Michell, R. (1983). Ca^{2+} and protein kinase C: two synergistic cellular signals. Trends in Biochemical Sciences, 8, 263-5.

Mohr, H. (1982). Pattern specification and realisation in photomorphogenesis. In: Encyclopaedia of Plant Physiology, 16, eds. W. Shropshire & H. Mohr, pp. 336-57. Berlin: Springer-Verlag.

Moreau, M., Guerrier, P. & Doree, M. (1976). Induction of meiosis by injection of heterologous protein kinase and phosphorylase kinase in Xenopus laevis oocytes (1). Journal of Experimental Zoology, 197, 435-41.

Nakajima, H., Yokota, T., Matsumoto, Nobuchi, M. & Takahashi, N. (1979). Relationship between hormone content and autonomy in various autonomous tobacco cells cultured in suspension. Plant and Cell Physiology, 20, 1489-99.

Nurse, P. (1980). Cell cycle control - both deterministic and probabilistic, Nature, 286, 9-10.

Osterreider, W., Brum, G., Hescheler, J., Trautwein, W., Flockerzi, V. & Hofmann, F. (1982). Injection of sub-units of cyclic AMP dependent protein kinase into cardiac myocytes modulates Ca^{2+} current. Nature, 298, 576-8.

Perry, T.O. (1971). Dormancy of trees in winter. Science, 171, 29-36.

Philips, I.R., Shepherd, E.A., Stein, J.L., Kleinsmith, L.J. & Stein, G.G. (1979). Nuclear protein kinase activities during the cell cycle of HeLa S_2 cells. Biochimica et Biophysica Acta, 565, 326-46.

Phillips, S.G. (1972). Repopulation of the Postmitotic Nucleolus by Preformed RNA. Journal of Cell Biology, 53, 611-23.

Polya, G.M. & Davies, J.R. (1982). Resolution of Ca^{2+} calmodulin activated protein kinase of wheat germ. FEBS Letters, 150, 167-71.

Pringle, J.H. & Hartwell, L.H. (1981). Genetic analysis of the cell cycle in Schizosaccharomyces pombe. In: The Molecular Biology of the Yeast, Saccharomyces, ed. J.N. Strathern, pp. 97-148. Cold Spring Harbor Laboratory.

Ranjeva, R., Refeno, G., Boudet, A.M. & Marme, D. (1983). Activation of plant quinate NAD 3-oxidoreductase by Ca^{2+} and calmodulin. Proceedings of the National Academy of Sciences, USA, 80, 5222-4.

Rao, M.V.N. & Prescott, D.M. (1970). Inclusion of Predivision Labelled Nuclear RNA in Post Division Nuclei in Amoeba proteus. Experimental Cell Research, 62, 286-92.

Roux, S.J. & Slocum, R.D. (1982). Role of calcium in mediating cellular functions important for growth and development in higher plants. In: Calcium and Cell Function, Vol. 3, ed. W.Y. Cheung, pp.409-53. New York: Academic Press.

Sembdner, G., Gross, D., Webisca, H.W. & Schneider, G. (1980). Biosynthesis of growth substances. In: Encyclopaedia of Plant Physiology, Vol. 9, ed. J. McMillan, pp. 281-444. Berlin: Springer-Verlag.

Setterfield, G. (1963). Growth regulation in excised slices of Jerusalem Artichoke Tuber tissue. Symposia of the Society for Experimental Biology, 17, 98-126.

Sheldrake, A.R. (1973). The production of hormones in higher plants. Biological Reviews, 48, 509-61.

Sikorska, M., MacManus, J.P., Walker, P.R. & Whitfield, J.F. (1980). The protein kinases of rat liver nuclei. Biochemical and Biophysical Research Communications, 93, 1196-203.
Sinnott, E.W. (1960). Plant Morphogenesis. New York: McGraw Hill.
Smith, J.A. & Martin, L. (1973). Do cells cycle? Proceedings of the National Academy of Sciences, USA, 70, 1263-7.
Steeves, T.A. & Sussex, I.M. (1972). Patterns in Plant Development. New Jersey: Englewoods Cliffs.
Steffensen, D.M. (1968). Shoot apical cell destinies in Zea mays. American Journal of Botany, 55, 354-69.
Stewart, R.N. (1978). Ontogeny of the primary body in chimeral forms of higher plants. Symposia of the Society for Developmental Biology, 36, 131-59.
Stratton, B.R. & Trewavas, A.J. (1981). Phosphorylation of histone H1 during the cell cycle of artichoke. Plant, Cell and Environment, 4, 419-26.
Street, H.E. (1968). Sterile root culture. In: Root Growth, ed. W.J. Whittington, pp. 18-42. New York: Plenum.
Tran Thanh Van, K.M. (1981). Control of morphogenesis in in vitro cultures. Annual Review of Plant Physiology, 32, 291-313.
Tran Thanh Van, K.M., Chlyah, H. & Chlyah, A. (1974). Regulation of organogenesis in thin layer of epidermal and sub-epidermal cells. In: Tissue Culture and Plant Science, ed. H.E. Street, pp. 101-41. London: Academic Press.
Tran Thanh Van, K.M. & Trinh, H. (1978). Morphogenesis in thin cell layers. In: Frontiers of Plant Tissue Culture 1978, ed. T.A. Thorpe, pp. 37-49. Calgary: International Association for Plant Tissue Culture.
Trewavas, A.J. (1976a). Post translational modification of proteins by phosphorylation. Annual Review of Plant Physiology, 27, 349-74.
Trewavas, A.J. (1976b). Plant growth substances. In Molecular Aspects of Gene Expression in Plants, ed. J.A. Bryant, pp. 249-98, London & New York: Academic Press.
Trewavas, A.J. (1979a). What is the function of protein phosphorylation in the plant nucleus? NATO FEBS Advanced Study Institute: Genome organisation and expression in plants. Abstract A22.
Trewavas, A.J. (1979b). Nuclear phosphoproteins in germinating cereal embryos and their relationship to the control of mRNA synthesis and the onset of cell division. Phytochemical Society Symposia, 16, 175-95.
Trewavas, A.J. (1981). How do plant growth substances work? Plant, Cell and Environment, 4, 203-28.
Trewavas, A.J. (1982a). Possible control points of cell development. In: The Molecular Biology of Plant Development, eds. H. Smith & D. Grierson, pp. 8-28. Oxford: Blackwell Scientific Publications.
Trewavas, A.J. (1982b). Growth substance sensitivity: the limiting factor in plant development. Physiologia Plantarum, 55, 60-72.
Trewavas, A.J. (1982c). The regulation of development and its relation to growth substances. What's New In Plant Physiology, 13, 41-4.
Trewavas, A.J. (1983a). Nitrate as a plant hormone. In: British Plant Growth Regulator Group Monographs 9, 97-110.
Trewavas, A.J. (1983b). Is plant development regulated by changes in the concentration of growth substances or by changes in the sensitivity to growth substances? Trends in Biochemical Science, 8, 354-7.
Wareing, P.F. (1979). What is the basis of stability of apical meristems. In: British Plant Growth Regulator Group Monographs 3, 1-11.

Weiler, E. & Spanier, K. (1981). Phytohormones in the formation of crown gall tumours. Planta, 153, 326-37.

REGULATION OF THE CELL DIVISION CYCLE IN CULTURED PLANT CELLS

M.W. Bayliss

INTRODUCTION

The growth of plant cells in culture, particularly the growth of single cells or small aggregates of cells in liquid medium of defined composition, provides the ideal environment in which to study control of the cell division cycle. Large homogenous populations of cells can, for example, be subjected to controlled alterations in nutrient and growth factor concentrations and easily subsampled to provide independent estimates of cell number doubling times and cell cycle phase parameters (e.g. Gould, Bayliss & Street, 1974; Bayliss, 1975, 1977a).

The development of plant tissue culture techniques has been intimately associated with the discovery and characterisation of the plant growth regulators auxin and cytokinin (Skoog & Miller, 1957) and in large part this has been due to the role of these hormones in permitting sustained cell division and cell expansion in the absence of culture differentiation. It is axiomatic that sustained growth in culture, no less than in the whole plant, requires continued cell division. The periodic assertion by some physiologists that plants grow merely by cell expansion has served only to hamper study of the central role of cell division (e.g. discussion by Brown, 1976; Haber & Foard, 1964).

The general topic of cell cycle in plant cell cultures has been reviewed several times recently (Gould, 1984; King, 1980a, b; Davidson, Aitchison & Yeoman, 1976). These reviews adequately describe cell cycle events in plant cell cultures in general and with texts such as Aherne, Camplejohn & Wright (1977) and Mitchison (1971) provide a complete description of the techniques used to measure cell cycle phase proportions. In his recent review of the control of the cell cycle in cultured plant cells, Gould (1984) has explored a number of definitions of the concept of "control". I propose in this text to use the term more narrowly and consider only the role of growth regulators and nutrient components of the culture medium, both in the initiation of growth *in vitro* and in the regulation of cell cycle events in established cell lines. This is, in part, an attempt to compare knowledge of events in plant cells with the extensive understanding of the role of hormones and nutrients in cultured animal cells (e.g. Stiles, Cochran & Scher, 1981). It is also, potentially, the most useful approach if we are to exploit our understanding of cell cycle control in plants through controlled modifications of crop plant performance, either by applications of growth regulatory chemicals or by alteration to the plant genotype.

INITIATION OF GROWTH IN CULTURE

Tissue explants

All plant cell cultures originate from the initiation of disorganised callus growth, usually on tissue or organ explants (Yeoman, 1973). Although some callus formation occurs normally as a wound response in plant tissues, sustained growth requires a culture medium providing both nutrients and growth regulators. In all cases, a supply of auxin is required to initiate callus growth, but requirements for an exogenous supply of cytokinin vary widely with species (e.g. Bayliss & Dunn, 1979). Although the gross characteristics of culture initiation have been described for a wide range of species, only in a very few cases have detailed cell cycle studies been performed.

The classical objects for such studies have been explants from pith tissue of tobacco (*Nicotiana tabaccum*), tuber tissues of artichoke (*Helianthus tuberosus*) and root tissues of pea seedlings (*Pisum sativum*). In all cases the explants come from quiescent non-dividing tissues, and culture conditions re-initiate cell division during the process of callus formation.

The simplest system is provided by the artichoke tuber explants, which consist almost entirely of G1-arrested diploid nuclei with a 2*C* DNA content (Yeoman & Mitchell, 1970). Addition of an auxin to the culture medium produces a synchronous progression of a proportion of the tuber cells through S-phase, G2 and mitosis (see Fig. 1). It has also been shown more recently that additions of abscisic acid increase the proportion of nuclei entering S-phase in this system (Minocha, 1979; Minocha & Halperin, 1974). Although the precise response in artichoke tuber explants depends on the condition of the tuber (e.g. Bennici, Cionini, Gennai & Cionini, 1982), this tissue provides fairly clear evidence for the action of auxin on a specific G1 cell cycle control point.

In contrast, results obtained with the tobacco pith explant system are much more difficult to interpret. Freshly isolated pith explants show a wound response leading to initiation of DNA synthesis (as shown by tritiated thymidine incorporation) even in basal media, and auxin and cytokinin additions separately or together enhance this response (Simard, 1971); Patau, Das & Skoog, 1957). In tissues allowed to age in culture however, only auxin additions lead to DNA synthesis (Simard, 1971). In no case have cytokinin additions alone led to mitosis, whether or not DNA synthesis was induced. Auxin additions by contrast produced both DNA synthesis and mitosis and were synergised by cytokinin (Das, Patau & Skoog, 1956, 1958; Patau *et al.*, 1957.

The complicating factor in pith tissue is the presence *in vivo* of a polyploid series of nuclei up to at least 8*C* (Das *et al.*, 1958). It

Fig. 1. Distributions of nuclear DNA contents in cultured artichoke tuber explants (A) before and (B) 24 hours after addition of 10^{-6}M 2,4-D. (Modified from Yeoman & Mitchell, 1970).

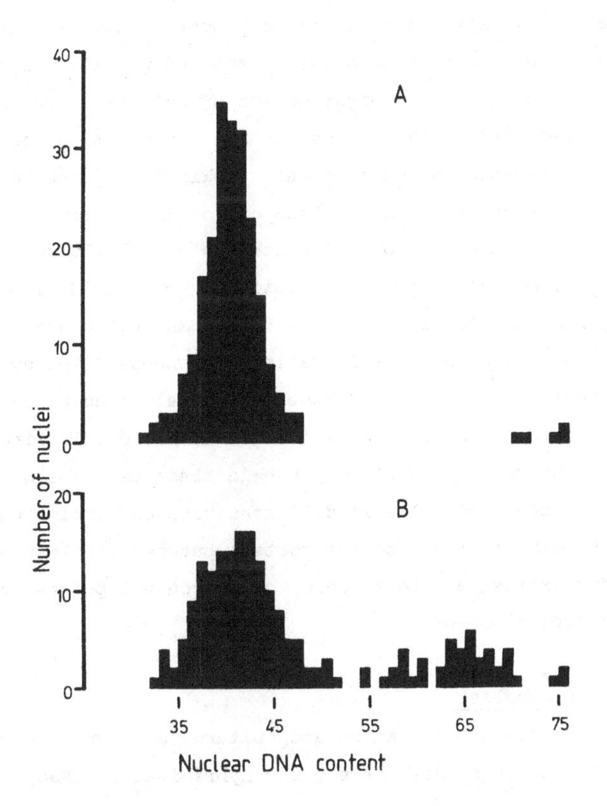

has therefore been difficult to ascertain the proportions of pith cells arrested in G1 and G2 (see p.171). However by treating explants with auxin and cytokinin in the presence of tritiated thymidine, Patau & Das (1961) have shown, using autoradiography, that mitoses occur in diploid and tetraploid cells both with and without preceding DNA replication. That is, cells arrested in both G1 and G2 can re-enter the division cycle in culture. As measurements of this type have not been made after separate auxin or cytokinin treatments, the tobacco pith system provides no evidence for the action of these hormones on specific cell cycle phases.

Root tissues of pea similarly contain both diploid and tetraploid nuclei _in vivo_ (see discussion by Matthyse & Torrey, 1969) and when cultured on media containing auxin and cytokinin, cells of both ploidy levels are stimulated into division (Torrey, 1965; Van't Hof & MacMillan, 1969). On media containing auxin alone, use of tritiated thymidine labelling and autoradiography has demonstrated that initial diploid mitoses result from cells arrested both in G1 and G2. Tetraploid cells were only stimulated into division by the addition of cytokinins, but in this case, the majority of initial tetraploid mitoses contained diplochromosomes and were unlabelled, i.e. cells which had arrested after an endoreduplication cycle were stimulated into division. After a longer period of culture in the presence of cytokinin, labelled tetraploid mitoses without diplochromosomes were visible, suggesting that prolonged exposure to cytokinin supported continuing DNA synthesis in tetraploid cells. In support of this, Matthyse & Torrey (1967) have shown that inhibition of DNA synthesis with 5-fluorodeoxyuridine (FUdR) prevents the appearance of both diploid and tetraploid mitosis in media containing kinetin. The pea variety "Little Marvel" contains very few tetraploid cortical cells and in this variety, Libbenga & Torrey (1973) have shown that culture in the presence of kinetin actively induces endoreduplication, leading to the appearance of a dividing tetraploid cell lineage in culture.

The complexity of response in these pea explants is in part related to response of cells in different tissues within the root (tetraploid cells being confined to the cortex) but nevertheless provides little evidence for exclusive effects on G1 or G2 control points by auxins or cytokinins respectively.

Protoplasts

Protoplast isolation and culture techniques were originally developed to permit studies of somatic hybridisation (Kao, 1980) or to allow rapid development of single cell clones from plants. To this end, protoplast isolation and culture conditions have now been described for a large number of plant species (Binding et al., 1981). Protoplasts also have a number of potential advantages over tissue explants for studies on initiation of cell cycle events in culture (Meyer & Aspart, 1983) though this facet has been little studied.

Protoplasts are usually isolated from leaf mesophyll cells, simply because these provide a convenient source of large numbers of identical cells. In tobacco (Nicotiana tabaccum), the species most frequently used for such studies, mesophyll cells are non-meristematic and probably arrest exclusively in G1. Galbraith, Mauch & Shields (1981) have used microfluorimetric measurements of nuclear DNA content on freshly isolated protoplasts to show a single peak of DNA content. Unfortunately, Galbraith et al., (1981) report no measurement of recognizable mitotic stages in diploid cells to fix the G1 and G2 DNA contents. However, they assume their measurements indicate a G1-arrested cell population in the mature mesophyll.

Culture of isolated protoplasts in the presence of auxin and cytokinin leads apparently to a semi-synchronous progression through S, G2 and mitosis (Cooke & Meyer, 1981; Zelcer & Galun, 1976) although cell cycle phase characteristics have not been determined by conventional techniques. It would appear that both auxin and cytokinin are required for the initiation of DNA synthesis and mitosis although the cytokinin requirement for S-phase initiation is only apparent after an initial culture period in the presence of auxin alone (Cooke & Meyer, 1981).

More attention has been devoted to studies of the interrelationship of cell wall formation and nuclear division. Using the specific inhibitor of wall formation, 2,6-dichlorobenzonitrile, Meyer & Herth (1978) and Galbraith & Shields (1982) were able to show that DNA synthesis and mitosis could proceed without cytokinesis to produce multinucleate cells. Other, less specific, inhibitors of wall formation such as coumarin and cellulase prevent both mitosis and cytokinesis (Schilde-Rentschler, 1977; Meyer & Herth, 1978).

The unique potential for somatic cell fusion offered by plant protoplasts has been used by Szabados & Dudits (1980) to study fusion of protoplasts containing mitotic and interphase nuclei. As had been shown earlier in similar fusions between animal cells (Johnson & Rao, 1970) the plant cell fusions induced premature chromosome condensation in the interphase nuclei. As this effect occurred in intergeneric fusions between

Triticum monococcum and Oryza sativa it provides evidence in plants for a generalized inducer of chromosome condensation.

THE ROLE OF MEDIUM CONSTITUENTS IN THE GROWTH OF ESTABLISHED CELL LINES

Auxins

In general, all plant cell cultures require an exogenous supply of auxin for continued undifferentiated growth. Omission of auxin leads either to a rapid cessation of growth and cell division (e.g. Gamburg, 1982; Volfova, Opatrni, Khvoika & Stoinova, 1982; Everett, Wang, Gould & Street, 1981; Codron et al., 1979; Nishi, Kate, Takahasi & Yoshida, 1977; King, 1976) or in suitable cultures to organogenesis or embryogenesis (see reviews by Kohlenbach, 1976; Reinert, 1973). In those cases where an auxin supply is not required for continued undifferentiated growth, elevated levels of endogenous auxin production have been implicated, often resulting from transformation by Agrobacterium tumefaciens (Morris et al., 1982; Meins, 1982; Weiler, 1981; Nakajima et al., 1979).

In those examples where auxin (usually supplied as the synthetic auxin 2,4-dichlorophenoxyacetic acid, 2,4-D) removal leads to cessation of cell division, it is unclear whether cells accumulate at any specific cell cycle phase. Everett et al., (1981) reported that there was no shift in nuclear DNA content distribution in Acer cultures starved of 2,4-D although their data (Fig. 2) are suggestive of a G1 accumulation. However, there was no evidence that re-supply of 2,4-D could produce the synchronous growth characteristic of a cell population arrested at a single point in the cell cycle. Re-supply of 2,4-D to starved cultures in fact produced only periodic oscillations of mitotic index whilst cell number increased exponentially.

Gamburg (1982) has described a somewhat similar phenomenon in cultures of tobacco. Using tritiated thymidine to label S-phase nuclei during naphthalene acetic acid (NAA)-stimulated culture growth, Gamburg showed initial mitoses in unlabelled nuclei, suggesting these cells had arrested in G2 during the preceding NAA starvation. Subsequently, labelled mitoses resulting from cells in G1 at the time of starvation, were evident in the culture. This evidence for cells blocked both in G1 and G2 during auxin starvation could lead to semi-synchronous growth on re-supply of auxin, and Gamburg's data (1982: Fig. 2) provide some evidence for this.

Nishi et al., (1977) claim to have synchronised cell suspension cultures of carrot by removal and re-supply of 2,4-D. No direct measurement of cell-cycle parameters was made and these authors conclude that 2,4-D starvation caused G1 arrest largely on the basis that a peak of mitotic activity occurred only after a period of DNA synthesis in cultures stimulated into growth by re-supply of 2,4-D. As no autoradiographic data were presented to verify that mitosis occurred only in cells which had undergone DNA synthesis after 2,4-D addition, (cf discussion of Gamburg's data above), and the reported mitotic index data suggested only a poor degree of synchrony, it remains to be proved that 2,4-D starvation caused exclusive arrest in G1 in these carrot cultures. At best, the results of Nishi _et al_. (1977) show only that 2,4-D starvation is very unlikely to have caused cell cycle arrest exclusively in G2. As exponential carrot cultures normally contain only approximately 9% of cells in G2 (Bayliss, 1975) cell arrest through 2,4-D starvation in the G1:G2 proportions typical of exponential cultures could equally have produced the results shown.

In the carrot cultures described by Bayliss (1975), removal of

Fig. 2. Distributions of nuclear DNA contents in cell suspension cultures of sycamore (A) growing exponentially in the presence of 2,4-D; (B) growth arrested in minus 2,4-D medium. (Modified from Everett _et al_., 1981).

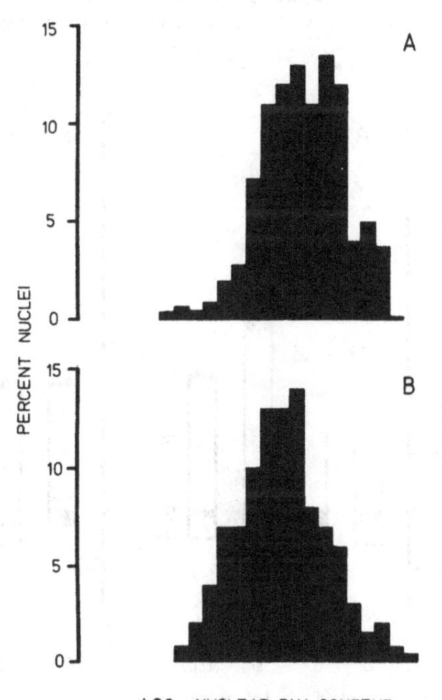

2,4-D led either to embryogenesis, or to a reduced growth rate, depending on the culture line used (Fig. 3a). In the embryogenic culture, removal of 2,4-D produced a marked reduction in cell cycle time, concomitant with the onset of morphogenesis. Cell cycle time was shortened largely through a reduction in the length of G1. Similar evidence for enhanced rates of mitosis and DNA synthesis during embryogenesis in carrot cultures has also been provided by Fujimura, Komamine & Matsumoto (1980) and in cultures if Pimpinella anisum by Huber, Constabel & Gamborg (1978).

In the partially auxin autotrophic carrot cell line described by Bayliss (1975) removal of 2,4-D extended cell cycle duration largely by an increase in duration of G1 (Fig. 4a). For both embryogenic and non-embryogenic cell lines, removal of 2,4-D produced increased G2 durations. In a further study of elevated 2,4-D levels, Bayliss (1977a) again demonstrated that increased cell cycle durations in carrot suspension cultures were attributable largely to extensions of G1, but also of G2 (Fig. 3b).

Fig. 3. (A) Cell cycle durations and phase proportions for 2 cell lines of carrot grown in the presence and absence of 2,4-D. (Modified from Bayliss, 1975). (B) Cell cycle durations and phase properties for a cell line of carrot grown at various 2,4-D concentrations. These values have been calculated from the data of Bayliss (1977a) on the assumption of a growth fraction of 1.0 at all 2,4-D concentrations.

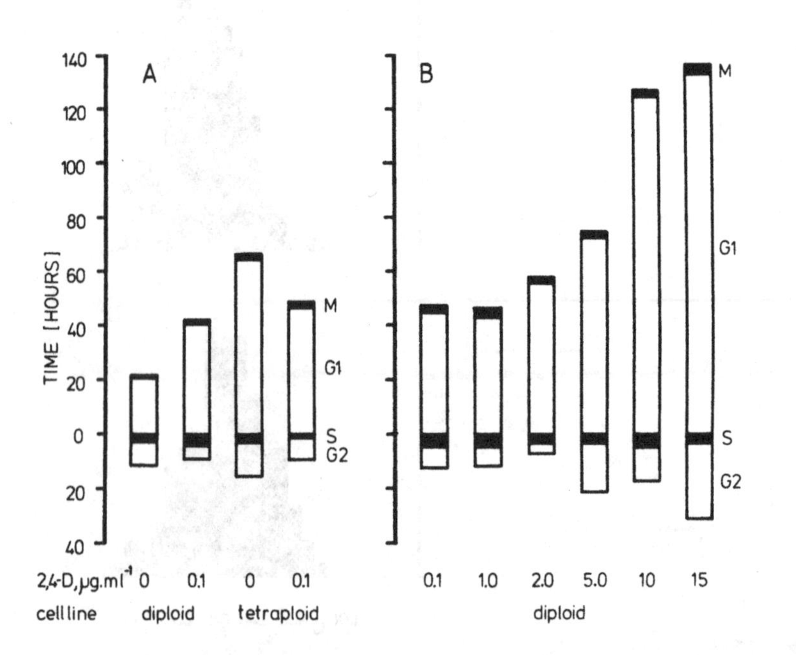

Cytokinins

Since their discovery, cytokinins have been heavily implicated in the control of cell division in cultured plant cells (Skoog & Armstrong, 1970; Fosket, 1977). Despite this, there have been very few attempts to define the precise role of these compounds in control of cell events.

Working with a cytokinin-requiring cell line of soybean, Wang, Everett, Gould & Street (1981) have shown that although transfer of cells to cytokinin-deficient medium prevented cell number increase, there was no obvious accumulation of cells with either G1 or G2 DNA contents. Re-supply of cytokinin produced oscillations of mitotic index but no evidence of a high degree of cell synchrony from either cell number or mitotic index data.

Similar effects of cytokinin starvation and re-supply on mitotic index have been shown previously for suspension cultures of tobacco by Joanneau (1971) and Joanneau & Tandeau de Marsac (1973). Despite mitotic

Fig. 4. (A) Cell number growth curves for 50 cm^3 batch suspension cultures of Paul's Scarlet Rose grown in the presence of 0(●); 1.0(O); 7.5(■) or 15 (■) μg ml^{-1} PP333. (B) Proportions of cells in S-phase (measured as frequency of labelled nuclei in autoradiographs of cells flash-labelled with tritiated thymidine) and mitosis, in 50 cm^3 batch suspension cultures of Paul's Scarlet Rose grown in control medium or medium containing 20 μg ml^{-1} PP333. (Previously unpublished observations of M.W. Bayliss).

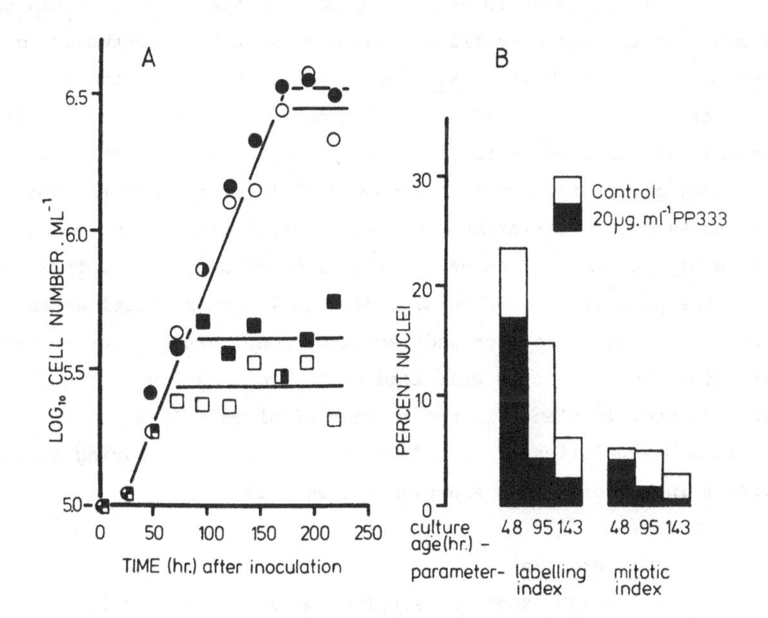

index oscillations on cytokinin re-supply, Joanneau & Tandeau de Marsac failed to detect any effect of cytokinin regime on DNA synthesis (as detected by tritiated thymidine incorporation) and inhibition of DNA synthesis by FUdR did not prevent mitotic index oscillations. Although a low frequency of mitoses occurred in the absence of cytokinin, these showed abnormal mitotic phase proportions indicative of abnormalities of spindle function.

In a series of papers on soybean cell suspension cultures, Fosket and his co-workers (reviewed by Fosket, 1977) have obtained similar results to Joanneau. Cytokinin starvation of the soybean cultures rapidly reduced mitotic index to zero, whilst DNA synthesis (as determined by tritiated thymidine incorporation) was unaffected. DNA extraction and thymidine incorporation data supported the idea that in the absence of cytokinin, cells were capable of successive rounds of DNA replication in the absence of mitosis i.e. the cells became polyploid. Some suggestion of this effect may also be visible in the nuclear DNA content histograms of Wang *et al.* (1981).

The only complete cell cycle analysis available for cytokinin treated cultures has unfortunately been performed on a cytokinin autotrophic strain of tobacco (Vyskot & Bezdek, 1982). In this system, kinetin addition produced an increased rate of thymidine incorporation into DNA despite previous evidence using the fraction labelled mitoses technique (Aherne *et al.*, 1977) which showed kinetin had no effect on cell cycle phase durations.

There seems to be quite good evidence that cytokinin effects are mediated through specific alterations in protein synthesis (discussed by Fosket, 1977 and Wang *et al.*, 1981) and in reviewing the topic, Fosket has concluded that cytokinins are required for the synthesis of G2 specific proteins essential for mitosis. It is perhaps more correct to view cytokinins as factors permitting normal centromere division and mitotic spindle formation, as evidenced for example by the abnormal mitoses in tobacco in the absence of cytokinin (Joanneau & Tandeau de Marsac, 1973). Thus cytokinins permit the partitioning of chromatids into daughter nuclei as an alternative to cycles of endomitosis or endoreduplication. This of course does not imply that cytokinin starvation will lead to G2 accumulation as the G2-G1 transition can occur in plants in the absence of mitosis (see p.170). Cytokinins have also been implicated in endoreduplication events during initiation of division in cultured tissue explants (see p.170).

Gibberellins

Although exogenous supplies of gibberellin promote the growth of cultures from low density inoculations (Stuart & Street, 1971) and can increase final cell numbers in culture (Street, Collins, Short & Simpkin, 1969; Digby, Thomas & Wareing, 1964) there are no reports of gibberellin-dependent culture growth. In consequence, there are no published cell cycle studies of the effects of gibberellins *in vitro*. As plant cell cultures are thought to be largely self-sufficient for gibberellins (see Digby *et al*., 1964) one approach to their study is the use of inhibitors of gibberellin biosynthesis.

The gibberellin biosynthesis inhibitor PP333 (paclobutrazole) acts by specifically blocking the oxidation of kaurene to kaurenoic acid both in fungi (J. Macmillan, unpublished data) and higher plants (Hildebrandt, Graebe, Rademacher & Jung, 1982; Dalziel & Lawrence, 1984). When added to suspension cultures of Paul's Scarlet Rose, PP333 prevents cell number increase after a time dependent on inhibitor concentration (Fig. 4a). The compound does not however prevent cell volume or cell dry weight increases, and inhibited cultures have greatly enlarged cells. Measurements of mitotic and labelling indices in the presence of inhibitory concentrations of PP333 showed rapid reductions in numbers of mitotic and S-phase nuclei in treated cultures (Fig. 4b). Examination of nuclear DNA content distributions by microspectrophotometry showed no evidence for specific cell cycle phase accumulation, though there was some evidence for the appearance of higher ploidy cells after treatment (Fig. 5). Essentially similar results have been obtained with the gibberellin biosynthesis inhibitors 'AMO' and 'Phosphon D' (J. Hall, unpublished data). This suggests that gibberellins are necessary for the normal sequence of DNA synthesis and mitosis but provides no evidence for a unique gibberellin dependent cell cycle control point.

Other growth regulators and nutrients

No detailed cell cycle studies in established cultures have been carried out with the other classes of plant growth regulators. Constable, Kurz, Chatson & Kirkpatrick (1977) succeeded in partially synchronizing cultures of soybean by periodically flushing with ethylene. The presence of quite sharp mitotic index peaks (maximum mitotic index 16%) suggests a high degree of synchrony was obtained, but unfortunately no observations of cell cycle parameters other than mitosis were made. Ethylene is generally regarded as an inhibitor of cell division (Goodwin, 1978) and Constabel *et al*. (1977) provide some evidence for cell cycle arrest at two points in response to ethylene.

Abscisic acid (ABA) is inhibitory to the growth of established cultures of Paul's Scarlet Rose (M.W. Bayliss, unpublished data) although at low concentrations, the compound has been reported to stimulate the incorporation of ^{32}P-phosphate into nucleic acids (Neskovic, Petrovic, Radojevic & Vujicic, 1977). No detailed cell cycle analyses of ABA effects have been reported.

Observations on the role of nutrient components of culture media on cell cycle progression have been more interesting. Bayliss & Gould (1974)

Fig. 5. Nuclear DNA content distributions for 50 cm^3 batch suspension cultures of Paul's Scarlet Rose during exponential growth (A) or stationary phase (B) in control medium and growth inhibited in medium containing 20 μg ml^{-1} PP333 (C). The bimodal stationary phase control distribution represents the presence of sub-populations of cells with chromosome numbers of 2n = 35 and 2n = 70. (Previously unpublished observations of M.W. Bayliss).

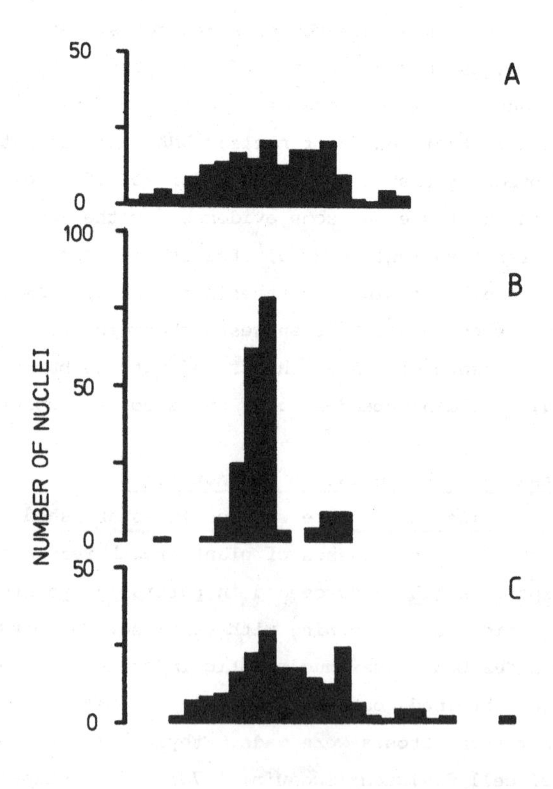

and Gould *et al.* (1974) have shown that cultured cells of sycamore (*Acer pseudoplatanus*) accumulate in G1 when starved of nitrogen in late stationary phase. Release from this starvation by transfer to fresh medium then leads to a series of synchronous divisions (Gould & Street, 1975; King, Cox, Fowler & Street, 1974). Modifications to the medium to produce starvation for phosphate or sucrose by contrast suggested arrest of cells in both G1 and G2 in proportions similar to those present in exponential cultures (Gould *et al.*, 1981). By contrast, Komamine, Morigaki & Fujimura (1978) obtained synchronisation of suspension cultures of *Vinca* after phosphate starvation and re-feeding. Species differences in response to nutrient starvation probably reflect selection of cell lines in media differing in limiting nutrients (Bayliss, 1977b). It would appear then that nutrient limitations more accurately identify cell cycle phase-specific control points in established cultures than do plant growth regulator treatments (see Trewavas, this volume).

DISCUSSION AND CONCLUSIONS

Work with animal cells, both *in vivo* and *in vitro*, supports the idea that non-dividing cell populations arrest in the G1 phase of the cell cycle, and that slow division rates are associated with increased G1 durations (Stiles *et al.*, 1981; Yanishevsky & Stein, 1981). Re-initiation of division activity in quiescent cell populations is then associated primarily with entry into S-phase. Both entry into prolonged G1 (or G0) phase and re-initiation of division are controlled by specific hormonal or nutrient factors present for example, in the serum component of culture media (Stiles *et al.*, 1981).

Simple models of this type can be discerned in the literature discussed in previous sections. Initiation of division in the artichoke tuber system (p.158) and probably in mesophyll protoplasts (p.160) largely under the influence of auxin, seems to be through movement of cells from G1 to S-phase. Extension of G1 in slowly growing cultures of *Acer* (Gould *et al.*, 1981) and carrot (Bayliss, 1977b) and G1 accumulation after nitrate (Gould *et al.*, 1981) or phosphate (Komamine *et al.*, 1978) starvation similarly highlight a major cell-cycle control point in G1.

This picture is complicated however, by the lack of clear-cut evidence that auxins and other growth regulators influencing cell division, act solely on a G1 control point. Auxin starvation certainly prevents cell division in established cultures (see p.162) but does not apparently lead to

G1 accumulation (Everett *et al.*, 1981). Data from all systems in which auxin starvation experiments have been performed, are consistent with cells arresting in the G1;G2 proportions found in dividing cultures. Re-feeding with auxin then produces periodic oscillation of mitotic index which may reflect synchronous cycles from these two cohorts of cells. Gamburg's observations using autoradiography (Gamburg, 1982) certainly demonstrates division of cells blocked in G2.

It is unfortunate that very few authors have tried to provide a precise definition of the state of cells apparently blocked in G2. In most instances, G2 has been characterised purely in terms of a 4*C* nuclear DNA content. However, the true definition of G2 must also include a description of the state of kinetochore replication and subsequent behaviour of cells on re-activation of the division cycle. The majority of differentiated plant tissues become endopolyploid through the processes of endomitosis (Tschermak-Woess, 1971) or endoreduplication (D'Amato, 1965) leading to the mixoploid state seen in tobacco pith and pea root explants (see p.166 and Nagl, Pohl & Radler, this volume). This implies that many cells with a 4*C* DNA content may not be functionally in a G2 state (see Fig. 6). Thus 4*C* cells may be in G1 after endomitosis, in which case they will have a tetraploid number of kinetochores, or be destined to undergo a second

Fig. 6. Chromosome configurations and DNA contents during normal and amitotic cell cycles in plant cells.

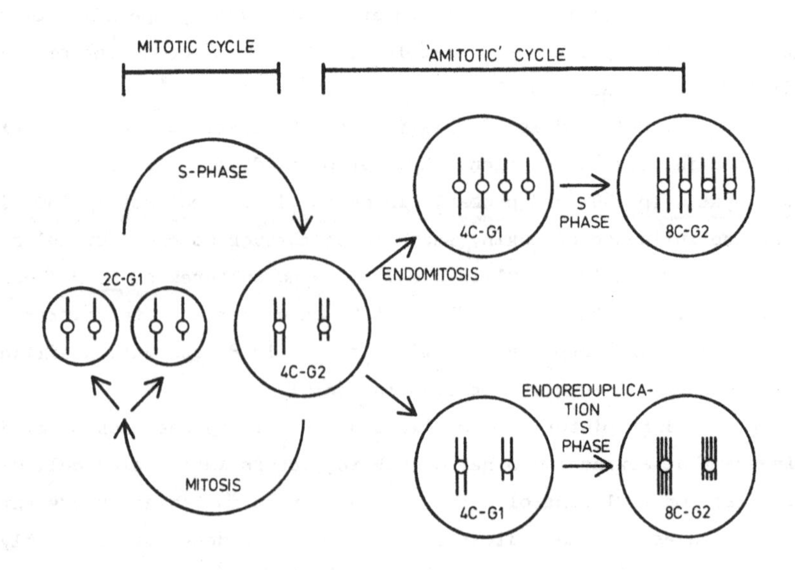

round of DNA synthesis with complete omission of mitosis, in which case they will still have the diploid number of kinetochores, but each with 4 chromatids (Fig. 6). Apparent arrest in G2, as evidenced by a 4C DNA content peak (e.g. Everett et al., 1981) thus may disguise a cell population arrested exclusively in G1 from a functional viewpoint.

Artificial induction of division in culture may break these amitotic cell cycles to produce for example diplochromosome-containing mitoses (Torrey, 1965; Patau & Das, 1961) or true tetraploid cell lineages (Libbenga & Torrey, 1973).

Many of the equivocal results obtained with cytokinins may be due to the involvement of these compounds in the switch between normal and amitotic cell cycles. Much of the evidence from cytokinin starvation experiments (see p.166) suggests that these compounds are necessary for normal mitosis, and cultures grown in the presence of auxin but without cytokinin undergo endoreduplication of endomitosis. This would explain the failure of Wang et al. (1981) to find evidence for specific cell cycle arrest after cytokinin starvation. Interestingly, tissue explants responsive to cytokinins (see p.160) are endopolyploid, which may suggest that lack of endogenous cytokinins in vivo can lead to cycles of endoreduplication or endomitosis. Mitotic index oscillations after re-supply of cytokinins in both tissue explants (see p.160) and established cultures (see p.165) could then reflect different cell cycle kinetics in diploid and polyploid cells rather than relief from specific cell cycle blocks.

The general justification for cell cycle studies in culture is that they can provide evidence for control systems in vivo. In the very few instances where a direct comparison is possible (Table 1) it is apparent that cell cycle times in culture are extended, largely through extension of G1. This in part reflects the very different cell morphology of meristematic cells in vivo and typical cultured cells (Gould, 1984), a feature borne out by immunological characterisation of cell types in culture (Raff, Hutchins, Knox & Clarke, 1978). Evidence of reduction in cell cycle times as cultured cells differentiate in the absence of hormones (see p.162) suggests that levels of auxin necessary to suppress organisation are inhibitory to cell cycle progression (see also Bayliss, 1977a) and that during the process of differentiation other factors affecting cell cycle progression may be more important (Albersheim et al., 1981; Meyer & Aspart, 1983).

In conclusion, the main factor limiting our ability to analyse factors controlling cell cycle progression in culture, is a lack of appropriate data (see also comments by Gould, 1984). Despite this, the results discussed above are compatible with the presence of a major control point in G1 which is fairly specifically modulated by both auxins and nutrients. This putative controlling role for nutrients has interesting parallels in those systems thought to govern apical dominance in plants (McIntyre, 1977). It is much less clear whether or not there is a true G2 control point as most studies have failed to provide all the necessary evidence to define functional G2 arrest. Cytokinin involvement with a putative G2 control point seems to be concerned with the switch between normal and amitotic cell cycles rather than as an absolute requirement for cell cycle progression. These arguments have interesting implications for the role of trigonelline in the promotion of G2 arrest *in vivo* (Evans & Tramontano, 1981; see Van't Hof, this volume). Although other growth regulators seem to influence cell cycle events (see p. 167) there are really too little data to provide a clear analysis of their role. Similarly, there is as yet insufficient information to decide whether the results obtained in culture provide evidence for the controls operating in the intact plant.

Table 1. Cell cycle phase durations in root tip cells and cell cultures of 3 plant species

Species	Cycle phase duration (h)					Reference
	T	G1	S	G2	M	
Daucus carota root tip	7.5	1.3	2.7	2.9	0.6	Bayliss, 1975
Daucus carota cell culture	51.2	39.6	3.0	6.2	2.4	Bayliss, 1975
Haplopappus gracilis root tip	10.5	3.5	4.0	1.4	1.6	Sparvoli, Gay & Kaufman, 1966
Haplopappus gracilis cell culture	22.0	9.3	6.4	4.9	1.4	Ericksson, 1967
Zea mays root tip	9.9	1.7	5.0	2.1	1.1	Verma, 1980
Zea mays cell culture	37.0	23.9	7.1	3.9	2.1	Gould, 1984

Albersheim, P., Darvill, A.C., McNeil, M., Valent, B.S., Sharp, J.K., Hahn, M.G., Lyon, G.D., Hodgson, B. & Woodward, M.D. (1981). Complex carbohydrates function as regulatory molecules within plants and as specific messages between plants and their pests. Heredity, 47, 158.

Aherne, W.A., Camplejohn, R.S. & Wright, N.A. (1977). An introduction to cell population kinetics. London: Edward Arnold.

Bayliss, M.W. (1975). The duration of the cell cycle of Daucus carota in vivo and in vitro. Experimental Cell Research, 92, 31-8.

Bayliss, M.W. (1977a). The effects of 2,4-D on growth and mitosis in suspension cultures of Daucus carota. Plant Science Letters, 8, 99-103.

Bayliss, M.W. (1977b). The causes of competition between two cell lines of Daucus carota in mixed cultures. Protoplasma, 92, 117-27.

Bayliss, M.W. & Dunn, S.D.M. (1979). Factors affecting callus formation from embryos of barley (Hordeum vulgare). Plant Science Letters, 14, 311-6.

Bayliss, M.W. & Gould, A.R. (1974). Studies on the growth in culture of plant cells XVIII. Nuclear cytology of Acer pseudoplatanus suspension cultures. Journal of Experimental Botany, 25, 772-83.

Bennici, A., Cionini, P.G., Gennai, D., & Cionini, G. (1982). Cell cycle in Helianthus tuberosus tuber tissue in relation to dormancy. Protoplasma, 112, 133-7.

Binding, H., Nehls, R., Kock, R., Finger, J. & Mordhurst, G. (1981). Comparative studies on protoplast regeneration in herbaceous species of the class dicotyledonae. Zeitschrift für Pflanzenphysiologie, 101, 119-30.

Brown, R. (1976). The Significance of Division in the Higher Plant, in Cell division in Higher Plants. Ed. M.M. Yeoman, pp. 3-46. London and New York: Academic Press.

Codron, H., Latche, A., Pech, J.C., Nebie, B. & Fallot, J. (1979). Control of quiescence and viability in auxin-deprived pear Pyrus-communis cultivar passecrassane cells in batch and continuous culture. Plant Science Letters, 17, 29-36.

Constable, F., Kurz, W.G.W., Chatson, K.B. & Kirkpatrick, J.W. (1977). Partial synchrony in soybean cell suspension cultures induced by ethylene. Experimental Cell Research, 105, 263-8.

Cooke, R. & Meyer, Y. (1981). Hormonal control of tobacco protoplast nucleic acid metabolism during in vitro culture. Planta, 152, 1-7.

Dalziel, J. & Lawrence, D.K. (1984). Biochemical and biological effects of kaurene oxidase inhibitors, such as paclobutrazol. British Plant Growth Regulator Group Monograph, in press.

D'Amato, F. (1965). Endopolyploidy as a factor in plant tissue development. Proc. Int. Conf. Plant Tissue Cult., Penn. State Univ., 1963. Ed. P.R. White, A.R. Grove, pp. 449-62. Berkeley, California: McCutchan.

Das, N.K., Patau, K. & Skoog, F. (1956). Initiation of mitosis and cell division by kinetin and indoleacetic acid in excised tobacco pith tissue. Physiologia Plantarum, 9, 640-51.

Das, N.K., Patau, K. & Skoog, F. (1958). Autoradiographic and microspectrophotometric studies of DNA synthesis in excised tobacco pith tissue. Chromosoma, 9, 606-17.

Davidson, A.W., Aitchison, P.A. & Yeoman, M.M. (1976). Disorganized systems, in Cell division in Higher plants. Ed. M.M. Yeoman, pp. 407-34. London & New York: Academic Press.

Digby, J., Thomas, T.H. & Wareing, P.F. (1964). Promotion of cell division in tissue cultures by gibberellic acid. Nature, 203, 547-8.

Eriksson, T. (1967). Duration of the mitotic cycle in cell cultures of Haplopappus gracilis. Physiologia Plantarum, 20, 348-54.

Evans, L.S. & Tramontano, W.A. (1981). Is trigonelline a plant hormone? American Journal of Botany, 68, 1282-9.

Everett, N.P., Wang, T.L., Gould, A.R. & Street, H.E. (1981). Studies on the control of the cell-cycle in cultured plant-cells. 2. Effects of 2,4-dichlorophenoxyacetic acid, 2,4-D. Protoplasma, 106,15-22.

Fosket, D.E. (1977). Regulation of the cell cycle by cytokinin, in Mechanism and control of cell division. Ed. T.L. Rost & E.M. Gifford. pp. 62-91. Stroudsberg, Pennsylvania: Dowden, Hutchinson & Ross.

Fujimura, T., Komamine, A. & Matsumoto, H. (1980). Aspects of DNA, RNA and protein synthesis during somatic embryogenesis in a carrot Daucus carota cultivar kurodagosun cell suspension culture. Physiologia Plantarum, 49, 255-60.

Galbraith, D.W. & Shields, B.A. (1982). The effects of inhibitors of cell wall synthesis on tobacco Nicotiana tabaccum cultivar xanthi protoplast development. Physiologia Plantarum, 55, 25-30.

Galbraith, D.W., Mauch, T.J. & Shields, B.A. (1981). Analysis of the initial stages of plant protoplast development using Hoechst 33258 reactivation of the cell cycle. Physiologia Plantarum, 51, 380-6.

Gamburg, K.Z. (1982). Regulation of cell division by auxin in isolated cultures, in Plant Growth Substances 1982, Ed. P.F. Wareing, pp. 59-67. London: Academic Press.

Goodwin, P.B. (1978). Phytohormones and growth and development of organs of the vegetative plant, in Phytohormones and related compounds: a comprehensive treatise, vol. II. Eds. D.S. Letham, P.B. Goodwin & T.J.V. Higgins. pp. 31-144. Amsterdam: Elsevier.

Gould, A.R. (1984). Control of the cell cycle in cultured plant cells. Boca Raton, Florida: C.R.C. press.

Gould, A.R., Bayliss, M.W. & Street, H.E. (1974). Studies on the growth in culture of plant cells XVII. Analysis of the cell cycle of asynchronously dividing Acer pseudoplatanus in cells in suspension culture. Journal of Experimental Botany, 25, 468-78.

Gould, A.R., Everett, N.P., Wang, T.L. & Street, H.E. (1981). Control of the cell cycle in cultured plant cells. I. Effects of nutrient limitation and nutrient starvation. Protoplasma, 106, 1-14.

Gould, A.R. & Street, H.E. (1975). Kinetic aspects of synchrony in suspension cultures of Acer pseudoplatanus. Journal of Cell Science, 17, 337-48.

Haber, A.H. & Foard, D.E. (1964). Further studies of gamma-irradiated wheat and their relevance to use of mitotic inhibition for developmental studies. American Journal of Botany, 51, 151-9.

Hildebrandt, E., Graebe, J.E., Rademacher, W. & Jung, J. (1982). Mode of action of new potent growth retardants: BAS 106 and Triazole compounds. Poster Abstract 762, 11th Int. conf. plant growth substs. Aberystwyth, U.K.

Huber, J., Constabel, F. & Gamborg, O.L. (1978). A cell counting procedure applied to embryogenesis in cell suspension cultures of anise, Pimpinella anisum. Plant Science Letters, 12, 209-16.

Joanneau, J.P. & Tandeau de Marsac, N. (1973). Stepwise effects of cytokinin activity upon tobacco cell division. Experimental Cell Research, 77, 167-74.

Joanneau, J.P. (1971). Controle par les cytokinines de la synchronisation des mitoses dans les cellules de tabac. Experimental Cell Research, 67, 329-37.

Johnson, R.T. & Rao, P.N. (1970). Mammalian cell fusion: induction of premature chromosome condensation in interphase nuclei. Nature, 226, 717-22.

Kao, K.N. (1980). Expression of foreign genetic material through protoplast fusion, through uptake of prokaryotic cells and cell organelles, in Plant Cell cultures: results and perspectives. Ed. F. Sala, B. Parisi, R. Cella & O. Ciferri. pp. 195-205. Amsterdam: Elsevier.

King, P.J. (1976). Studies on the growth in culture of plant cells, part 20, utilization of 2,4-D by steady-state cell cultures of Acer pseudoplatanus. Journal of Experimental Botany, 27, 1053-72.

King, P.J. (1980a). Plant tissue culture and the cell cycle. Advances in Biochemical Engineering, 18, 1-38.

King, P.J. (1980b). Cell proliferation and growth in suspension cultures. International Review of Cytology suppl. IIA, 25-53.

King, P.J., Cox, B.J., Fowler, M.W. & Street, H.E. (1974). Metabolic events in synchronized cell cultures of Acer pseudoplatanus. Planta, 117, 109-22.

Kohlenbach, H.W. (1976). Basic aspects of differentiation and plant regeneration from cell and tissue culture, in Plant tissue culture and its biotechnological application, Eds. W. Barz, E. Reinhard & M.H. Zenk. pp. 355-366. Berlin: Springer-Verlag.

Komamine, A., Morigaki, T. & Fujimura, T. (1978). Metabolism in synchronous growth and differentiation in plant tissue and cell cultures, in Frontiers of plant tissue culture 1978. Ed. T.A. Thorpe, pp. 159. Calgary Int. Assoc. Plant Tiss. Cult.

Libbenga, K.R. & Torrey, J.G. (1973). Hormone induced endoreduplication prior to mitosis in cultured pea root cells. American Journal of Botany, 60, 293-9.

Matthyse, A.G. & Torrey, J.G. (1967). DNA synthesis in relation to polyploid mitoses in excised pea root segments cultured in vitro. Experimental Cell Research, 48, 484-98.

Matthyse, A.G. & Torrey, J.G. (1969). Factors limiting the stimulation of polyploid mitoses in intact pea roots and excised root segments. Botanical Gazette, 130, 62-9.

McIntyre, G.I. (1977). The role of nutrition in apical dominance, Symposia of the Society for Experimental Biology, 31, 251-74.

Meins, F. (1982). Habituation of cultured plant cells, in Molecular biology of plant tumours. Ed. G. Kahl & J.S. Schell. pp. 3-32. New York: Academic Press.

Meyer, Y. & Aspart, L. (1983). The first mitotic cycle of mesophyll protoplasts, Experientia suppl., 46, 93-100.

Meyer, Y. & Herth, W. (1978). Chemical inhibition of cell wall formation and cytokinesis but not of nuclear division in protoplasts of Nicotiana tabaccum cultivated in vitro. Planta, 142, 253-62.

Minocha, S.C. (1979). Abscisic acid promotion of cell division and DNA synthesis in Jerusalem-artichoke Helianthus tuberosus tuber tissue cultured in vitro. Zeitschrift für Pflanzenphysiologie, 92, 327-40.

Minocha, S.C. & Halperin, W. (1974). Hormones and metabolites which control tracheid differentiation, with or without concomitant effects on growth, in cultured tuber tissue of Helianthus tuberosus L. Planta, 116, 319-31.

Mitchison, J.M. (1971). The biology of the cell cycle. Cambridge: Cambridge University Press.

Morris, R.O., Akiyoshi, D.E., MacDonald, E.M.S., Morris, J.W., Regier, D.A. & Zaerr, J.B. (1982). Cytokinin metabolism in relation to tumour induction, in Plant growth substances 1982. Ed. P.F. Wareing. pp. 175-84. London: Academic Press.

Nakajima, H., Yokota, T., Matsumoto, T., Noguch, M. & Takahashi, N. (1979). Relationship between hormone content and autonomy in various autonomous tobacco Nicotiana tabaccum cells cultured in suspension. Plant and Cell Physiology, 20, 1489-1500.

Neskovic, M., Petrovic, J., Radojevic, L. & Vujicic, R. (1977). Stimulation of growth and nucleic acid biosynthesis at low concentration of abscisic acid in tissue culture of Spinacia oleracea. Physiologia Plantarum, 39, 148-54.

Nishi, A., Kate, K., Takahasi, M. & Yoshida, R. (1977). Partial synchronization of carrot cell culture by auxin deprivation. Physiologia Plantarum, 39, 9-12.

Patau, K. & Das, N.K. (1961). The relationship of DNA synthesis and mitosis in tobacco pith tissue cultured in vitro. Chromosoma, 11, 553-72.

Patau, K., Das, N.K. & Skoog, F. (1957). Induction of DNA synthesis by kinetin and indoleacetic acid in excised tobacco pith tissue. Physiologia Plantarum, 10, 949-66.

Raff, J.W., Hutchins, J.F., Knox, R.B. & Clarke, A.E. (1978). Cell recognition - antigenic determinants of plant organs and their cultured callus cells. Differentiation, 12, 179-86.

Reinert, J. (1973). Aspects of organization - organogenesis and embryogenesis, in Plant tissue and cell culture. Ed. H.E. Street. pp. 338-55. Oxford: Blackwell.

Schilde-Rentschler, L. (1977). Role of the cell wall in the ability of tobacco protoplasts to form callus. Planta, 135, 177-82.

Simard, A. (1971). Initiation of DNA synthesis by kinetin and experimental factors in tobacco pith tissues in vitro. Canadian Journal of Botany, 49, 1541-49.

Skoog, F.F. & Armstrong, D.J. (1970). Cytokinins. Annual Review of Plant Physiology, 21, 359-84.

Skoog, F. & Miller, C.O. (1957). Chemical regulation of growth and organ formation in plant tissues cultured in vitro. Symposia of the Society for Experimental Biology, 11. 118-31.

Sparvoli, E., Gay, H. & Kaufman, B.P. (1966). Duration of the mitotic cycle in Haplopappus gracilis. Caryologia, 19, 65-71.

Stiles, C.D., Cochran, B.H. & Scher, C.D. (1981). Regulation of the mammalian cell cycle by hormonés, in The Cell Cycle, Ed. P.C.L. John. pp. 119-139. Cambridge: Cambridge University Press.

Street, H.E., Collins, H.A., Short, K. & Simpkin, P. (1969). Hormonal control of cell division and expansion in suspension cultures of Acer pseudoplatanus L: the action of kinetin, in Proc. 6th Int. Conf. Plant Growth Subst. Ed. F. Wightman. pp. 489-504. Ottawa: Runge Press.

Stuart, R. & Street, H.E. (1971). Studies on the growth in culture of plant cells, 10. Further studies on the conditioning of culture media by suspensions of Acer pseudoplatanus cells. Journal of Experimental Botany, 22, 96-106.

Szabados, L. & Dudits, D. (1980). Fusion between interphase and mitotic plant protoplasts, induction of premature chromosome condensation. Experimental Cell Research, 127, 442-6.

Torrey, J.G. (1965). Cytological evidence of cell selection by plant tissue culture media. Proc. Int. Conf. Plant tissue cult. Penn State Univ. Eds. P.R. White & A.R. Grove. pp. 473-84. Berkeley, California: McCutchan.

Tschermak-Woess, E. (1971). Endomitosis, in Handbuch der Allgemeinein Pathologie. Ed. H.W. Altmann. pp. 569-625. Berlin: Springer-Verlag.

Van't Hof, J. & MacMillan, B. (1969). Cell population kinetics in callus tissues of cultured pea root segments. American Journal of Botany, 56, 42-51.

Verma, R.S. (1980). The duration of G1, S, G2 and mitosis at four different temperatures in Zea mays L. as measured with tritiated thymidine. Cytologia, 45, 327-33.

Volfova, A., Opatrni, Z., Khvoika, L. & Stoinova, E. (1982). Effect of nonindole auxin phenyl acetic-acid on the growth and ultra-structure of a tobacco cell strain. Fiziologia Rastenii (Sofia), 8, 19-29.

Vyskot, B. & Bezdek, M. (1982). An autoradiographic study of DNA synthesis in cytokinin autotrophic tobacco cells - thymidine and 5 bromo-2 deoxyuridine incorporation promoted by kinetin. Zeitschrift für Pflanzenphysiologie, 106, 431-46.

Wang, T.L., Everett, N.P., Gould, A.R. & Street, H.E. (1981). Studies on the control of the cell cycle in cultured plant cells. 3. The effects of cytokinin. Protoplasma, 106, 23-36.

Weiler, E.W. (1981). Dynamics of endogenous growth-regulators during the growth cycle of a hormone autotrophic plant cell culture. Naturwissenschaften, 68, 377-8.

Yanischevsky, R.M. & Stein, G.H. (1981). Regulation of the cell cycle in eukaryotic cells. International Review of Cytology, 69, 223-59.

Yeoman, M.M. (1973). Tissue (callus) cultures - techniques, in Plant tissue and cell culture. Ed. H.E. Street, pp. 31-8. Oxford: Blackwell.

Yeoman, M.M. & Mitchell, J.P. (1970). Changes accompanying the addition of 2,4-D to excised jerusalem artichoke tuber tissue. Annals of Botany, 34, 799-810.

Zelcer, A. & Galun, E. (1976). Culture of newly isolated tobacco protoplasts; precursor incorporation into protein, RNA and DNA. Plant Science Letters, 7, 331-6.

GENETIC AND EPIGENETIC CONTROL OF THE PLANT CELL CYCLE

T. Cavalier-Smith

INTRODUCTION

The timing and coordination of the events of the cell cycle depend both on genetic controls and environmental conditions; observed cell cycle properties depend on the interaction between them. The phrase 'epigenetic controls' was chosen partly to stress this and partly because the outcome of these interactions depends also on the developmental state of the individual cell; for example the cell cycle differs in meiotic cells (Bennett, 1976) and mitotic cells. Cell cycles are therefore not static units of development, but dynamic sets of processes that undergo developmental or epigenetic modifications. Limited space prevents consideration of all of these, so I have selected six specific problems for discussion:

(1) Determination of the overall pattern of cell growth during the cell cycle.
(2) The genetic determination of nuclear volume and its growth.
(3) How replication and division are coordinated with cell growth.
(4) The differences and similarities between binary and multiple fission cell cycles.
(5) The genetic control of cell volume.
(6) The control of the length of the cell cycle.

Studies of cell cycle mutants have been very few in plants (Howell & Naliboff, 1973; Howell, 1974; Warr & Quinn, 1977; Sato, 1976; T. Cavalier-Smith, unpublished data) compared with the yeasts *Saccharomyces* (Carter, 1981) and *Schizosaccharomyces* (Nurse & Fantes, 1981), which have also been more thoroughly studied physiologically. In this chapter I therefore compare the plant cell cycle with that of yeasts, and also with animals and protozoa, since it is increasingly clear that the fundamental features of the cell cycle are similar in plants and other eukaryotes.

THE DISTINCTION BETWEEN CELL GROWTH AND CELL DIVISION

Study of the mechanisms of cell growth during the plant cell cycle has been somewhat neglected in comparison with the growth of non-proliferating and differentiating plant cells that occurs by vacuolization and cell elongation. Thus, the whole cell cycle is frequently referred to as 'cell division' (Brown, 1976; Yeoman, 1976). But as cell division forms only part of the fundamental processes of the cell cycle, to use it also as a label for the whole cycle encourages neglect of the other processes, and may cause confusion. Though Brown (1976) criticised the distinction drawn by Sachs (1887) between growth and division, this distinction is the essential first step towards understanding the plant cell cycle.

It is true that in meristematic cells growth and division normally keep step with each other so as to maintain a steady state where average cell volume and mass are constant despite an increase in cell number. But the natural examples of cell growth without division and cell division without growth emphasised by Sachs, and the experimental separation of cell growth and division by differential inhibition, or in mutants, show that cell growth and division are in principle quite separate processes, and also that in 'typical' meristematic cell cycles where they alternate and maintain a steady state there must be specific mechanisms to coordinate them. Clarification of the mechanisms of growth itself, and of its coordination with replication and division, depends on clearly recognising this.

Cell growth in proliferating cells may be measured either as increase in cell volume or in mass. It depends on the integration of four basic processes:

1. Net transport of molecules into the cell.
2. Net biosynthesis of macromolecules such as RNA and protein and of small molecules such a nucleotides, amino acids and phospholipids.
3. Intracellular transport of newly made molecules into the appropriate cellular compartment, e.g. RNA from nucleus to cytoplasm, many proteins into chloroplasts and mitochondria, lipids from internal membranes to the plasma membrane.
4. Assembly of molecules to form macromolecular assemblies such as membranes, microtubules, microfilaments, cell walls.

These component processes can be and have been studied in their own right, but the specific problem of cell growth during the cell cycle is how they are integrated with each other so as to maintain a steady state over many cell generations.

LIMITS TO GROWTH

Fundamental to the problem of growth regulation is the question which, if any, of these four processes controls the cell growth rate? Is there a universal answer to this question for all plants, for all eukaryotes, or even for all cells? One approach is to study the temporal pattern of growth during the cell cycle. Is it stepwise or continuous? If continuous, is it linear, exponentially increasing, sigmoidal or some other pattern? This question has been studied in bacteria, yeasts, and protozoa (see references in Mitchison, 1971; Lloyd, Poole & Edwards, 1982; Elliott & McLaughlin, 1983). It is found that cells grow continuously during interphase, but the pattern of growth in cell volume may be linear, exponential or sigmoidal depending on the species. Cell volume and dry mass often show different patterns of increase; in consequence cell density changes regularly and reproducibly through the cell cycle. Cell protein content also frequently shows a different pattern of increase from cell volume, so that to treat it as a measure of cell size (Donnan & John, 1983) is unwise. Though some of the variations reported may result from technical errors, the overall evidence does not support the hypothesis of a universal pattern of cell growth (Kubitschek, 1971). This makes it all the more important to study the question in both uni- and multicellular plants and not to rely on extrapolation from other organisms.

THE PATTERN OF CELL VOLUME GROWTH DURING THE PLANT CELL CYCLE

Classical studies indicated that cell volume increased continuously during interphase both in higher plant meristems and in synchronous cultures of the unicellular algae *Chlorella* (Tamiya, 1963; John *et al.*, 1973) and *Chlamydomonas* (Kates, Chiang & Jones, 1968; Mihara & Hase, 1971). But distinction between linear and exponential patterns requires a precision better than 3%. We have therefore used electronic cell volume determinations to study the pattern of increase in synchronized cultures of the unicellular green alga *Chlamydomonas* (N. Harborne & T. Cavalier-Smith, unpublished data, Fig. 1). Three interesting conclusions emerge from our work: (1) the pattern of increase is clearly linear (2) larger cells appear to grow at a greater rate than smaller cells, as judged by the greater slope of the upper quartile and lower slope of the lower quartile compared with the median and (3) a rise in temperature causes an immediate increase in growth rate and a drop in light intensity causes an immediate decrease.

A basic feature of linear growth is that in an exponentially growing cell population the overall rate must effectively be doubled once every cell cycle (Mitchison, 1971); in light-dark synchronized Chlamydomonas this doubling in growth potential must be in the dark period. Linear growth has frequently been observed in bacteria; Kubitschek (1971) suggested that it results from a constant rate of uptake of nutrients resulting from a constancy in the number and activity of nutrient uptake sites which must be doubled once every cell cycle. However in bacteria, protein usually appears to be accumulated exponentially; thus the rate of protein synthesis is not

Fig. 1. Growth in cell volume during the 12h light period of light-dark synchronized Chlamydomonas reinhardii. Median cell volumes (O) and upper (O) and lower (□) quartiles of the cell volume distribution, were determined by Coulter counter and Channelyser on cells grown at 20°C as previously described (Craigie & Cavalier-Smith, 1982).

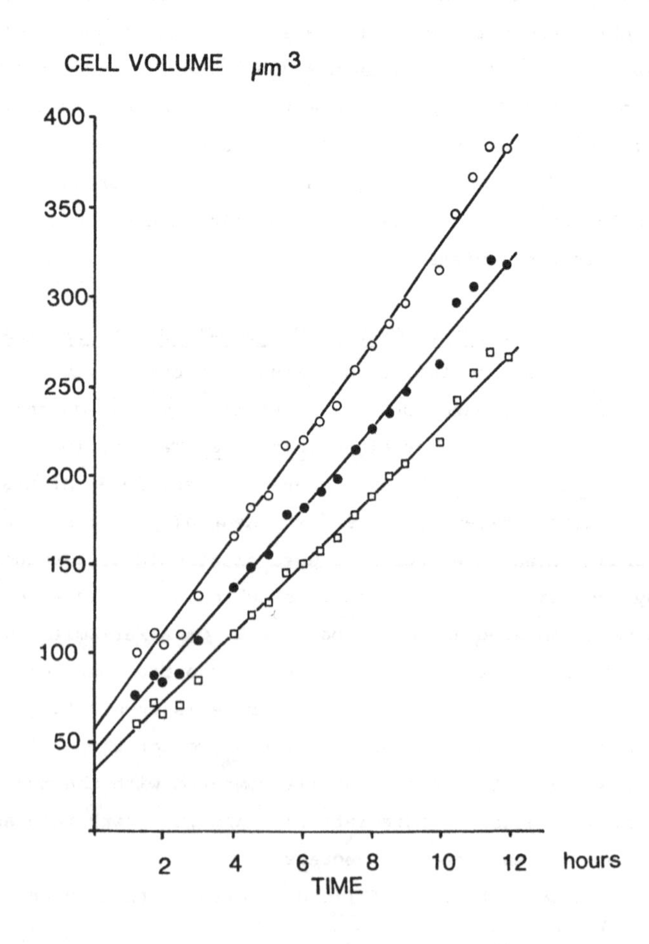

constant, as it would be if limited by a constant rate of nutrient uptake, but exponentially increasing; this suggests that it increases in direct proportion to the number of ribosomes and messenger RNA molecules and is therefore limited by them and not by nutrient uptake. This implies that nutrient *uptake* is not rate-limiting to growth, and that transport into the cell could only control cell volume if it did so by governing the uptake of osmotically active ions that were responsible for the osmotic control of cell volume as in non-growing animal cells. However the linear growth in volume of the rigid bacterial cell can more plausibly be explained in terms of a constant rate of wall assembly at a fixed number of sites, which are doubled once every cell cycle (Previc, 1970; Pritchard, 1974; Donachie, Jones & Teather, 1977).

This may also be true for unicellular plants such as *Chlamydomonas* and *Chlorella*. Since plant cell walls, unlike those of bacteria, grow as a result of exocytosis of vesicles containing wall precursors, I suggest that the constant growth rate results simply from the presence on the cell surface of a constant number of sites at which exocytosis can occur. Such sites might consist of rosettes of integral membrane proteins comparable to those at the sites where trichocysts undergo exocytosis in *Paramecium* (Beisson *et al.*, 1976). These sites may be partitioned between daughters by cytokinesis and new sites may be added to the plasma membrane only at this time, prior to the formation of new daughter cell walls. Variations in cell volume resulting from inaccurate partitioning of the mother cell into daughters will cause corresponding variations in the number of exocytosis sites. Cells that start their cell cycle larger than average will therefore have more exocytosis sites and be able to grow correspondingly faster. However the rate of wall growth should depend not only on the number of exocytosis sites but also on the number of vesicles fusing per site per unit time; this would depend on the temperature and local concentration of vesicles which in turn should increase with temperature and light intensity; at each light intensity and temperature a different steady state rate of wall growth would be set up for cells of a given initial volume and number of sites.

Linear growth in cell volume also occurs in animal cells, and in *Euglena* (Cook, 1961), (which I currently classify in the kingdom Protozoa rather than Plantae (Cavalier-Smith, 1983)). Though *Euglena* and animal cells lack a cell wall their plasma membrane presumably grows by exocytotic vesicle fusion, so the above explanation of linear growth may be widely applicable to eukaryotes.

Protein accumulates exponentially during the Chlorella cell cycle (John et al., 1982) and in Chlamydomonas (Lien & Knutsen, 1973); Donnan & John, 1983; Rollins, Harper & John, 1983). This supports my thesis that in these green algae, as in bacteria, the pattern of volume growth is controlled by the rate of wall assembly rather than protein synthesis. In the yeast Saccharomyces, protein increases exponentially (Elliott & McLaughlin, 1983), but volume growth is sigmoidal, and the cell density fluctuates regularly during the cell cycle as in Escherichia coli; that cell volume growth is controlled by exocytosis and not macro-molecular synthesis is clearly shown by mutants defective in exocytosis that fail to grow in volume despite continuing to synthesise protein exponentially and accumulating so much product that the cells become exceptionally dense (Schekman, 1982). I proposed (Cavalier-Smith, 1975), that actin microfilaments are essential for exocytosis; recent evidence on the distribution of such filaments in mutant and wild type yeasts during the cell cycle (Adams & Pringle, 1984) suggests that their location may control the sites of exocytosis for wall growth. Thus the assembly of the cell cortex and/or wall may be the basic determinant of the pattern of growth during the cell cycle in all eukaryotes, but that its temporal control is not the same in all species; for example if exocytosis sites were inserted continuously into the plasma membrane an exponential pattern of volume increase should occur, as observed during the first three quarters of the cell cycle in the fission yeast Schizosaccharomyces (Elliott & McLaughlin, 1982). The pattern of growth need not be the same, therefore, in higher plants, or even in all algae, as in Chlamydomonas.

GROWTH OF NUCLEAR VOLUME DURING THE CELL CYCLE

The discovery that nuclear volumes in animal tissues fell into a multimodal geometric series (Jacobj, 1925) was what led to the idea that chromosomes reproduced at a specific time in interphase (Hertwig, 1939); Swift (1953) subsequently showed that nuclear DNA content was doubled in interphase and correlated with nuclear volume. This correlation between nuclear volume and ploidy, together with the isometric correlation between meristematic nuclear volumes and DNA content in angiosperms (Price, Sparrow & Nauman, 1973), led to the 'skeletal DNA' hypothesis which suggested that total nuclear DNA content is the primary determinant of nuclear volume (Cavalier-Smith, 1978). However DNA content cannot be the sole determinant since nuclear volume varies at different times or in different tissues of the same species without corresponding changes in DNA content, but with corresponding variations in non-histone protein content (Alfert, Bern & Kahn, 1955; Bennett, 1974). The discovery of the proteinaceous nuclear lamina, underlying and attached to the nuclear envelope and pore complexes on the one hand and to parts of the nuclear DNA on the other, confirmed that proteins also play a role in the nuclear skeleton and the control of nuclear volume.

In animals the nuclear lamina proteins (lamins) are disassembled at prophase and dispersed throughout the cell when the nuclear envelope breaks down and reassembled at telophase on the surface of the still-condensed chromosomes (Gerace & Blobel, 1980; Jost & Johnson, 1981). According to my skeletal DNA hypothesis (Cavalier-Smith, 1978, 1980, 1982, 1984a,b) interphase nuclear volume is controlled by (1) the total amount of nuclear DNA (2) its pattern of folding and degree of compaction or dispersal and (3) its manner of attachment, to the nuclear lamina. Nuclear volume could therefore change during the cell cycle in response to changes in any of these factors. In most eukaryote cell types nuclear volume and dry mass increase predominantly during late S-phase and G2, i.e. following DNA replication, and therefore do not parallel the more uniform increase in volume of the whole cell (Lloyd _et al._, 1982; Schel, Steenbergen, Bekers & Wanka, 1978). This stepwise growth of the nucleus is consistent with my skeletal DNA hypothesis.

According to Woodward, Rasch & Swift (1961) and Lyndon (1967) nuclear volume in plant meristems increases during the cell cycle by a factor of five or six rather than a factor of two. One interpretation of this would be that a nuclear swelling and shrinkage cycle is superimposed on the basic two-fold growth pattern seen in other eukaryotes; there is electron microscopic evidence for a shrinkage phase in the closed nuclear cycle of the alga _Polytoma_ (Gaffal, Wolf & Schneider, 1983). But recent studies suggest that the heterogeneity in nuclear volume in meristems may be attributable mainly to systematic variation in nuclear volume from cell to cell rather than to shrinkage and swelling during individual cell cycles (Bansal & Davidson, 1978).

The possibility that growth in nuclear volume plays a role in the transition from G1 to S-phase is suggested by the finding that large nuclei initiate replication earlier than smaller nuclei in animals (Yen & Pardee, 1979), and that in plants the larger nucleus of two sisters divides first (Armstrong, 1983); moreover nuclear swelling precedes DNA replication when somatic nuclei are injected into animal eggs (Gurdon & Woodland, 1968) or when cells with inert nuclei are fused with actively replicating cells (Johnson & Rao, 1971). There is evidence that replication forks (Maul, 1982) and origins (Goldberg, Collier & Cassel, 1983) are attached to the nuclear matrix; and that in yeast and animal cells protein synthesis is a prerequisite for initiation of replication (Lloyd et al., 1982). It is reasonable to conclude that initiation of DNA replication depends either on the prior unfolding of the chromatin or on the prior synthesis of new nuclear matrix sites for the attachment of the replication machinery, or both.

COORDINATION BETWEEN CELL GROWTH AND DNA REPLICATION AND CELL DIVISION

For proliferating cells in a steady state, as in plant meristems or algal cultures, cells must on average divide once for every two-fold increase in volume. The oldest hypothesis concerning such control is that division is triggered directly by growth in cell volume: Hertwig (1903) suggested that as the cell grew the ratio nuclear volume/cytoplasmic volume (the karyoplasmic ratio) would steadily decline; its falling below a critical level would trigger nuclear growth and division. Popoff (1908-9) showed that in protozoa nuclear growth did indeed lag behind cell growth, and that cutting off some of the cytoplasm did delay division if it was done sufficiently early in the cell cycle, but that after a certain point the cell was committed to divide despite such experimental increases in the karyoplasmic ratio.

Demonstration that non-dividing eukaryote cells are usually blocked in G1 of the cell cycle rather than in G2 (Mitchison, 1971), together with classical studies of the cell cycle in bacteria (Helmstetter et al., 1968), led to the view (Cooper, 1979, 1982; Craigie & Cavalier-Smith, 1982) that the primary control point in the cell cycle is usually not the control of cell division but the control of DNA replication: in most cases initiation of replication leads automatically to division after a relatively constant period of preparations for division. However, phenomena such as endopolyploidy in eukaryotes (see Nagl, Pohl & Radler, this volume), where repeated rounds of DNA replication occur without mitosis, and the formation of non-septate filaments with many nucleoids in bacteria (Donachie, 1981) show that in both eukaryotes and prokaryotes division can in principle be controlled independently of replication; there are thus two major potential control points in the cell cycle: initiation of replication (the G1/S-phase transition) and initiation of division (the G2/M transition) (Van't Hof & Kovacs, 1972; Van't Hof, this volume).

The emphasis on potential control points should be noted; control need not be actually exerted at a potential control point. Suppose, for example, that passage past a particular cell cycle stage depends on the accumulation of a certain molecule to a particular level. If when that stage is reached too little of the molecule is present, the cell must await further accumulation before proceeding, so the controlling role of the molecule will be apparent; but if enough molecules had accumulated before the cell reached that stage they would no longer be limiting, and the cell could immediately proceed to the next stage: the action of the molecule would not be apparent, but cryptic. This is an important idea since it helps to explain the reasons for the differing cell cycle patterns in different species and at different developmental stages.

In *Schizosaccharomyces pombe* potential control points exist at both the G1/S and the G2/M boundaries, and attainment of a critical cell volume is necessary both for the initiation of replication and for initiation of mitosis (Fantes, 1977; Fantes & Nurse, 1977). However, under certain growth conditions the cell volume may already be above one or other of these critical levels so that either may be cryptic. Such volume-dependent controls are referred to as sizers; the evidence for them in yeast is very extensive (Mitchison, 1977; Fantes & Nurse, 1981; Fantes, 1983). Now that some of the genes apparently involved in the G1/S transition have been cloned (Beach, Durkacz & Nurse, 1982) it should eventually be possible to decide between the possible mechanisms for size control discussed elsewhere (Fantes *et al.*, 1975; Fantes & Nurse, 1981).

None of the theories for sizer control suggest that cell size *per se* controls the G1/S or G2/M transitions. They propose instead that size operates by affecting the concentration of molecules: one such model is the repressor dilution model (Pritchard, Barth & Collins, 1969) where a fixed amount of repressor is made once per cell cycle and gradually diluted by cell growth until its concentration falls below a certain threshold level, which will occur at a critical cell volume. Another model is the autorepressor model of Sompayrac & Maaløe (1973) where negative feedback keeps repressor concentration constant and thereby ensures that the total amount per cell of a transcriptionally coupled replication initiator molecule increases in proportion to cell volume.

Such molecular mechanisms for the sizer are fundamentally

statistical or probabilistic in nature, as they depend on random collisions between molecules: therefore size controls must be inherently 'sloppy' and subject to random variation, as is clearly the case in yeast (Wheals, 1980), and also to modulation by environmental or genetic factors that influence the concentration or activity of the interacting molecules. Authors studying mammalian cells, impressed by the great variability of G1 compared with S+G2, proposed a purely probabilistic model for the G1/S transition, in which cell size played no part (Smith & Martin, 1973). But evidence that larger mammalian cells do divide earlier than smaller ones led to the modification of the transition probability model by adding the idea of some kind of size control (Brooks, 1981). The simplest modification is to suppose that size operates simply by increasing the probability of the G1/S transition; this modified transition probability model differs only in name from a sloppy size control model; an alternative modification where size control and the random transition relate to two successive but distinct events in the cell cycle, is favoured by Brooks but seems unnecessarily complex. The key point, however, is that some degree of randomness and some degree of size control is necessary to explain the facts.

It has generally been assumed that sizer control over division is essential to ensure that cell volumes do not diverge indefinitely from the population mean as a result of random variations in growth rates and inequalities in cell division. However, Brooks (1981) argued that under certain conditions mean cell size could be maintained by a size-independent control of division. His two basic conditions were (1) that division probability is roughly proportional to growth rate, and (2) that the rate of cell volume increase is not proportional to cell volume, i.e. is not exponential. Our observations of cell growth rates in *Chlamydomonas* discussed above show that growth is linear and not exponential, so the second condition is satisfied. But we also showed that larger daughter cells do grow faster than smaller ones; therefore the first condition, that division probability is roughly proportional to growth rate implies for *Chlamydomonas* that division probability is not independent of cell size. Other organisms also show a positive correlation between cell size and growth rate (Fantes & Nurse, 1977; Jagadish, Lorincz & Carter, 1977; Tovey & Brouty-Boye, 1976). Brook's assumption that division probability can be independent of cell size, and yet proportional to cell growth rates appears to be invalid.

Donnan and John (1983) implied that larger cells divide earlier not because they are larger but because they grow faster. They suggest that 'division follows a timed period during which faster growing cells grow more'. They claim that in Chlamydomonas 'cycle duration is under timer control' and imply that sizers play no role in the timing of division. However they ignored our earlier findings (Craigie & Cavalier-Smith, 1982) that in light-dark synchronized cultures of Chlamydomonas large cells do undergo cytokinesis earlier than smaller ones. Thus the total length of the cell cycle from the onset of growth to cytokinesis is not constant, as it would be if it were 'under timer control', but variable (as their data in fact confirm), and its length is inversely related to cell size. In these respects Chlamydomonas is no different from other eukaryotes for which data are available. In higher plants also, some kind of sizer appears to be involved in the timing of cell division; in meristems the larger cell of two sisters divides first (Ivanov, 1971), its larger nucleus having been produced at the preceding telophase by the unequal distribution of RNA and protein at the mitotic spindle poles (Armstrong, 1983).

MULTIPLE FISSION CELL CYCLES

The vegetative life cycles of many algae, e.g. Chlamydomonas, Chlorella, Acetabularia, as well as protozoa (e.g. many Sporozoa), consist of a succession of multiple fission cell cycles (Cavalier-Smith, 1980). The cleavage divisions of the fertilised egg in animals and higher plants are also examples of multiple fission cell cycles, and share basic features with those of unicellular organisms, namely the temporal separation of growth and division. Cells dividing by binary fission grow two-fold on average between successive divisions; whereas with multiple fission growth is far more pronounced, the volume increasing many times in the absence of division; repeated divisions then occur in the absence of growth.

The simplest modification of cell cycle models to account for multiple fission cell cycles is to postulate a positive inhibition of the G1/S transition (Cavalier-Smith, 1980). For the duration of the inhibition the cell would continue to grow in the absence of DNA replication and division, as indeed is observed; when the block is relieved the normal G1/S sizer mechanism would operate, and, provided the cell had at least doubled in volume during the blockage, S-phase would be initiated and division occur. If during the block the cell had multiplied at least four-fold in volume, and accumulation of initiator or dilution of repressor had continued in proportion to cell growth, then a second round of replication and division would initiate immediately after the first division without further growth;

repeated replication and division would go on until the volume of the daughter cells fell below the threshold at which the sizer mechanism no longer permits initiation of replication. This model simply explains why multiple fission cell cycles lack a G1 period in the second and subsequent division cell cycles whether in animal cleavage (Mitchison, 1971) or in algae (Donnan & John, 1983). A G1 will occur in cell cycles only where daughter cells are smaller than the critical size for initiation of replication as determined by the G1/S sizer, or when the sizer is temporarily blocked.

What is the nature of the block, and how is it imposed and removed? If the central mechanism of the sizer is simply the binding of a soluble initiator or repressor to a G1/S controlling gene, such as one of the 'start' genes of yeast (Beach *et al.*, 1982), then the block could be imposed either by the binding of a separate controlling RNA or protein molecule to the 'start' gene, or by directly modifying the gene (for example by methylation, inversion, transposition, or nicking to alter its degree of supercoiling) in such a way as to interfere with the binding of the repressor/initiator. Relief of the block in the case of cleaving eggs would be triggered by fertilisation, and imposition of the block by an early step in differentiation of the animal oocyte or plant megaspore.

In the *Chlamydomonas* cell cycle, I propose that imposition of the block is controlled by a sizer. This can only be imposed when cell size is below a critical threshold; thus large *Chlamydomonas* mother cells will carry out successive replications and divisions until their daughters fall below this size as we (Craigie & Cavalier-Smith, 1982) and later Donnan & John (1983), previously proposed. The block to further replication will then be imposed and cells will be unable to replicate or divide until the block is removed. But what removes the block? It is possible that in light-dark synchronized *Chlamydomonas* removal of the block depends on a minimum period of illumination. Spudich & Sager (1980), Craigie & Cavalier-Smith (1982) and Donnan & John (1983) have all shown that a minimum period of illumination, commonly about eight hours, is needed to allow division. We have shown (Craigie & Cavalier-Smith, 1982 and unpublished) that cells subjected to less than the minimum period fail to divide even if they have more than doubled in size; yet at lower light intensities cells that have only doubled in size are able to divide once if they have been exposed to light for a long enough period.

Donnan & John (1983) transferred synchronised *Chlamydomonas* cells into the dark at successively later times during the light period; at early times, darkness blocked subsequent division, but ceased to do so just before the initiation of DNA replication. This fits my proposal that the light-dependent release of the block directly leads to replication. The minimum light period was very little affected by temperature (though it lengthened steadily as light intensity and consequently growth rate was reduced). They confusingly referred to it as being controlled by a 'timer', despite the fact that a timer (Fantes & Nurse, 1981) refers specifically to a fixed sequence of events (e.g. S+G2) that takes an approximately constant time at any _one_ temperature _despite widely varying growth rates_, but which may vary at _different_ temperatures. The minimum light period cannot be determined simply by the growth rate as they suggest, since _Chlamydomonas_ growth rates vary greatly with temperature as well as light intensity. I suggest instead that the rate of energy supply to the cell determines the minimum light period; in autotrophic cultures this will be affected by light intensity but not temperature, but if acetate were substituted for light, temperature should affect the uptake of acetate and therefore the timing of replication.

Donnan & John's (1983) observation, that the timing of DNA replication is the same in synchronous cultures established from small and from large cells, is compatible with our observation that larger cells undergo cytokinesis before smaller cells if the sizer for the G1 transition was cryptic under their growth conditions, and if the effect of cell volume that we observed affects the rate of progression from S to cytokinesis and not the timing of the initiation of S; the latter is in agreement with my present hypothesis of a blockage of the sizer mechanism during the light period of synchronous cultures, so long as the timing of the removal of the block is independent of cell volume.

Present evidence therefore is consistent with the idea that multiple fission cell cycles and binary fission cell cycles share a common potential control point at the G1/S transition that is a 'sloppy sizer' and that the difference between the two types of cycle depends on a reversible block imposed on the sizer mechanism in multiple fission cycles.

Although the primary potential control point is at the G1/S transition, control may, as discussed above, also potentially be exerted at the G2/M transition (see Van't Hof, this volume). In addition, the existence of developmental phases where nuclear division occurs in the absence of cyto-kinesis, as in gymnosperm female gametophytes and early embryos, shows that the transition from mitosis to cytokinesis is a third potential control point, and that the coupling between mitosis and cytokinesis is not automatic.

THE GENETIC DETERMINATION OF CELL VOLUME AND CELL CYCLE LENGTHS

The existence of the sizer-based primary control point at the G1/S phase transition also has important implications for the genetic control of cell volume and of cell cycle length, both of which vary widely in different species of plants. The mean volume of proliferating cells will be determined primarily by the setting of the G1/S sizer mechanism (Cavalier-Smith, 1980, 1982, 1984b); prior to differentiation, non-dividing cells will have the same volume as proliferating cells, but the genetic control of their final volume must depend on a specific block to cell growth, with respect to which they fall into three categories:

(1) if growth is blocked in G1 they will have the same ploidy and approximately the same cell volume as proliferating cells.

(2) if differentiation initially blocks only division then they will continue to grow to a much larger size than do proliferating cells and replication will be initiated by the G1/S sizer for every doubling in cell volume; large endopolyploid cells will automatically be produced (see Nagl, Radler & Pohl, this volume); the size and degree of endopolyploidy will be controlled by whatever eventually prevents further cell growth.

(3) if differentiation initially blocks only the G1/S sizer mechanism then large non-polyploid cells will be produced just as in the case of multiple fission cell cycles; their eventual size will be controlled by the cessation of growth as in endopolyploid cells.

Cell cycle length will be determined primarily by the cell growth rate. The much discussed positive correlations between cell volume and cell cycle length and genome size (Cavalier-Smith, 1978) are most simply explained as the indirect result of an isometric correlation between nuclear volume and cell volume and between nuclear volume and genome size, as discussed in detail elsewhere in relation to my skeletal DNA hypothesis (Cavlier-Smith, 1984a). Similarly the correlation between the length of S-phase and genome size (Francis, Kidd & Bennett, this volume) may perhaps be explained, not in terms of a direct control by genome size of the length of S-phase, but as the indirect result of the correlation between genome size and cell volume, and between cell volume and cell growth rates; the varying length of S-phase in different species, and at different developmental stages in one species, can be explained simply in terms of selection for the most economic number of replisomes (replication complexes) to ensure DNA replication in the available time, which itself depends primarily on the cell growth rate (Cavalier-Smith, 1984b). However, it must be stressed that it is difficult to extrapolate theories based on uni-cellular systems phased by light and dark transitions to multicellular tissue systems.

CELL GROWTH AND THE CONTROL OF CELL PROLIFERATION

Discussions of the control of cell proliferation in higher plants often assume that what is being controlled is either DNA replication or cell division. But if, as I have argued, these two processes are often controlled secondarily by changes in cell volume, the primary controls over plant cell cycles and rates of proliferation may often instead be caused by the inhibition or stimulation of cell growth, which could be quantitatively controlled in different regions of the plant body by the level of nutrients and hormones.

It used to be thought that cell growth depended on programmed successive gene activations (Mitchison, 1977), but recent studies of less perturbed synchronous cultures show that the hundreds of different commoner proteins made during the cell cycle are virtually all synthesised at the same exponential rate throughout in bacteria, yeasts (Elliott & McLaughlin, 1983), and in multiple fission plant cell cycles (John *et al.*, 1982; Rollins, Harper & John, 1983). Thus unlike replication and division that typically alternate during the cell cycle in a periodic fashion that is truly cyclic, the biosynthetic component of growth processes is non-cyclic and continuous.

Since cell growth depends on (a) nutrient transport into the cell (b) molecular biosynthesis and (c) organelle assembly, it could be controlled by regulating any of these three processes. The most economical point for such control in cells of multicellular plants would be by regulation of nutrient uptake; a primary limitation of nutrients would lead to secondary reductions in biosynthesis and consequently in assembly, lack of growth would prevent initiation of replication and division. Control of the spatial distribution of a limiting nutrient could, therefore, control the relative rates of cell proliferation within a meristem.

REFERENCES

Adams, A.E.M. & Pringle, J.R. (1984). Relationship of actin and tubulin distribution to bud growth in wild-type and morphogenetic-mutant *Saccharomyces cerevisiae*. *Journal of Cell Biology*, 98, 934-45.

Alfert, M., Bern, H.A. & Kahn, R.H. (1955). Hormonal influence on nuclear synthesis. IV. Karyometric and microspectrophotometric studies on rat thyroid nuclei in different functional states. Acta Anatomica, 23, 185-205.

Armstrong, S.W. (1983). Mitotic asymmetry: Differential behaviour of sister nuclei. Ph.D. Thesis, McMaster University.

Bansal, J. & Davidson, D. (1978). Heterogeneity of meristematic cells of *Vicia faba*. Caryologia, 31, 161-77.

Beach, D., Durkacz, B. & Nurse, P. (1982). Functionally homologous cell cycle control genes in budding and fission yeast. Nature, 300, 706-9.

Beisson, J., Lefort-Tran, M., Pouphile, M., Rossignol, M. & Satir, B. (1976). Genetic analysis of membrane differentiation in *Paramecium*. Freeze-fracture study of the trichocyst cycle in wild-type and mutant strains. Journal of Cell Biology, 69, 126-43.

Bennett, M.D. (1974). Nuclear characters in plants. Brookhaven Symposia on Biology, 25, 344-66.

Bennett. M.D. (1976). The cell in sporogenesis and spore development. In Cell Division in Higher Plants ed. M.M. Yeoman, pp. 161-98. London and New York: Academic Press.

Brooks, R.F. (1981). Variability in the cell cycle and the control of proliferation. In The Cell Cycle ed. P.C.L. John. pp. 35-61. Cambridge: Cambridge University Press.

Brown, R. (1976). Significance of Division in Higher Plants. In Cell Division in Higher Plants ed. M.M. Yeoman, pp. 3-46. London and New York: Academic Press.

Carter,B.L.A. (1981). The control of cell division in *Saccharomyces cerevisiae*. In The Cell Cycle. ed. P.C.L. John, pp. 99-117. Cambridge Cambridge University Press.

Cavalier-Smith, T. (1975). The origin of nuclei and of eukaryotic cells. Nature, 256, 463-8.

Cavalier-Smith, T. (1978). Nuclear volume control by nucleoskeletal DNA, selection for cell volume and cell growth rate, and the solution of the DNA C-value paradox. Journal of Cell Science, 34, 247-78.

Cavalier-Smith, T. (1980). *r*- and *K*-tactics in the evolution of protist developmental systems: cell and genome size, phenotype diversifying selection, and cell patterns. BioSystems, 12, 43-59.

Cavalier-Smith, T. (1982). Skeletal DNA and the evolution of genome size. Annual Review of Biophysics and Bioengineering, 11, 273-302.

Cavalier-Smith, T. (1983). A 6-kingdom classification and a unified phylogeny. In Endocytobiology II. ed. H.E.A. Schenk & W. Schwemmler. pp. 1027-34, Berlin: de Gruyter.

Cavalier-Smith, T. (1984a). Cell volume and the evolution of eukaryotic genome size. In The Evolution of Genome Size ed. T. Cavalier-Smith. In press. Chichester: Wiley.

Cavalier-Smith, T. (1984b). DNA replication and the evolution of genome size. In The Evolution of Genome Size ed. T. Cavalier-Smith. In press. Chichester: Wiley.

Cook, J.R. (1961). *Euglena gracilis* in synchronous division I. Dry mass and volume characteristics. Plant and Cell Physiology, 2, 199-202.

Cooper, S. (1979). A unifying model for the G1 period in prokaryotes and eukaryotes. Nature, 280, 17-9.

Cooper, S. (1982). The continuum model: application to G1-arrest and G(0). In Cell Growth. ed. C. Nicolini. pp. 315-36. New York: Plenum.

Craigie, R.A. & Cavalier-Smith, T. (1982). Cell volume and the control of the *Chlamydomonas* cell cycle. Journal of Cell Science, 54, 173-91.

Donachie, W.D. (1981). The cell cycle of Escherichia coli. In The Cell Cycle ed. P.C.L. John, pp. 63-83. Cambridge: Cambridge University Press.

Donachie, W.D., Jones, N.C. & Teather, R.M. (1977). The bacterial cell cycle. Symposia of the Society for General Microbiology, 23, 9-44.

Donnan, L. & John, P.C.L. (1983). Cell cycle control by timer and sizer in Chlamydomonas. Nature, 304, 630-3.

Elliott, S.G. & McLaughlin, C.S. (1983). The yeast cell cycle: coordination of growth and division rates. Progress in Nucleic Acid Research and Molecular Biology, 28, 143-76.

Fantes, P.A. (1977). Control of cell size and cycle time in Schizosaccharomyces pombe. Journal of Cell Science, 24, 51-67.

Fantes, P.A. (1983). Control of timing of cell cycle events in fission yeast by the wee 1^+ gene. Nature, 302, 153-5.

Fantes, P.A. & Nurse, P. (1977). Control of cell size at division in fission yeast by a growth-modulated size control over nuclear division. Experimental Cell Research, 107, 377-86.

Fantes, P.A. & Nurse, P. (1981). Division timing: controls, models and mechanisms. In The Cell Cycle ed. P.C.L. John. pp. 11-33. Cambridge: Cambridge University Press.

Fantes, P.A., Grant, W.D., Prithcard, R.H., Sudbery, P.E. & Wheals, A.E. (1975). The regulation of cell size and the control of mitosis. Journal of Theoretical Biology, 50, 213-44.

Gaffal, K.P., Wolf, K.W. & Schneider, G.J. (1983). Morphometric and chronobiological studies on the dynamics of the nuclear envelope and the nucleolus during mitosis of the colourless phytoflagellate Polytoma papillatum. Protoplasma, 118, 19-35.

Gerace, L. & Blobel, G. (1980). The nuclear envelope lamina is reversibly depolymerised during mitosis. Cell, 19, 277-87.

Goldberg, G.I., Collier, I. & Cassel, A. (1983). Specific DNA sequences associated with the nuclear matrix in synchronised mouse 3T3 cells. Proceedings of the National Academy of Sciences, USA, 80, 6887-91.

Gurdon, J.B. & Woodland, H.R. (1968). The cytoplasmic control of nuclear activity in animal development. Biological Reviews, 43, 233-67.

Helmstetter, C.E., Cooper, S., Pierucci, O. & Revelas, E. (1968). The bacterial life sequence. Cold Spring Harbor Symposia on Quantitative Biology, 33, 809-22.

Hertwig, R. (1903). Uber Korrelation von Zell- und Kerngrösse und ihre Bedeutung für die geschlechtliche Differenzierung und die Teilung der Zelle. Biologische Zentralblatt 23, 49-62, 108-19.

Hertwig, G. (1939). Abweichungen vom Verdoppelungsvolumen der Zellkerne und ihre Deutung. Anatomische Anziege, 87, 65-73.

Howell, S.H. (1974). An analysis of cell cycle controls in temperature-sensitive mutants of Chlamydomonas reinhardii. In Cell Cycle Controls, ed. G.M. Padilla, I.L. Cameron & A. Zimmerman. pp. 235-49. New York: Academic Press.

Howell, S.H. & Naliboff, J.A. (1973). Conditional mutants in Chlamydomonas reinhardii blocked in the vegetative cell cycle. I. An analysis of cell cycle block points. Journal of Cell Biology, 57, 760-72.

Ivanov, V.B. (1971). Critical size of the cell and its transition to division. I. Sequence of transition to mitosis for sister cells in the corn seedling root tip. Ontogenez 2, 524-35. (1972 English translation of Russian original by Consultants Bureau).

Jacobj, W. (1925). Uber das rhythmische Wachstum der Zellen durch Verdopplung ihres Volumens. Roux Archiv für Entwicklungs Mechanik, 106, 124-92.

Jagadish, M.N., Lorincz, A. & Carter, B.L.A. (1977). Cell size and cell division in yeast cultured at different growth rates. FEMS Microbiology Letters, 2, 235-7.

John, P.C.L., McCullough, W., Atkinson, A.W. Jr., Forde, N.G. & Gunning, B.E.S. (1973). The cell cycle in Chlorella. In The cell cycle in development and differentiation ed. M. Balls and F.S.Billett. pp. 61-76. Cambridge: Cambridge University Press.

John, P.C.L., Lambe, C.A., McGookin, R., Orr, B. & Rollins, M.J. (1982). Poly(A)$^+$ RNA populations, polypeptide synthesis and macromolecule accumulation in the cell cycle of the eukaryote Chlorella. Journal of Cell Science, 55, 51-67.

Johnson, R.T. & Rao, P.N. (1971). Nucleo-cytoplasmic interactions in the achievement of nuclear synchrony in DNA synthesis and mitosis in multinucleate cells. Biological Reviews, 46, 97-155.

Jost, E. & Johnson, R.T. (1981). Nuclear lamina assembly, synthesis and disaggregation during the cell cycle in synchronised HeLa cells. Journal of Cell Science, 47, 25-33.

Kates, J.R., Chiang, K.S. & Jones, R.F. (1968). Studies on DNA replication during synchronized vegetative growth and gametic differentiation in Chlamydomonas reinhardii. Experimental Cell Research, 49, 121-35.

Kubitschek, H.E. (1971). The distribution of cell generation times. Cell and Tissue Kinetics, 4, 113-22.

Lien, T. & Knutsen, G. (1973). Phosphate as a control factor in cell division of Chlamydomonas reinhardii, studied in synchronous culture. Experimental Cell Research, 78, 79-88.

Lloyd, D., Poole, R.K. & Edwards, S.W. (1982). The Cell Division Cycle: Temporal Organization and Control of Cellular Growth and Reproduction. London: Academic Press.

Lyndon, R.F. (1967). The growth of the nucleus in dividing and non-dividing cells of the pea root. Annals of Botany, 31, 133-46.

Maul, G.G. (1982) ed. The Nuclear Envelope and the Nuclear Matrix. New York: Liss.

Mihara, S. & Hase, E. (1971). Studies on the vegetative life cycle of Chlamydomonas reinhardii Dangeard in synchronous culture I. Some characteristics of the cell cycle. Plant and Cell Physiology, 12, 225-36.

Mitchison, J.M. (1971). The Biology of the Cell Cycle. Cambridge: Cambridge University Press.

Mitchison, J.M. (1977). The timing of cell cycle events. In Mitosis, Facts and Questions eds. M. Little, N. Paweletz, C. Petzelt, H. Postingl, D. Schroeter & H-P. Zimmerman. pp. 1-19. Berlin: Springer-Verlag.

Nurse, P. & Fantes, P.A. (1981). Cell cycle controls in fission yeast: a genetic analysis. In The Cell Cycle, ed. P.C.L. John. pp. 85-98. Cambridge: Cambridge University Press.

Popoff, M. (1908-9). Experimentelle Zell-studien. Archiv für experimental Zellforschung I. 245-379, II, IV.

Previc, E.P. (1970). Biochemical determination of bacterial morphology and the geometry of cell division. Journal of Theoretical Biology, 27, 471-97.

Price, H.J., Sparrow, A.H. & Nauman, A.F. (1973). Correlations between nuclear volume, cell volume and DNA content in meristematic cells of herbaceous angiosperms. Experientia, 29, 1028-9.

Pritchard, R.H. (1974). On the growth and form of the bacterial cell. Philosophical Transactions of the Royal Society of London, B. 267, 303-36.

Pritchard, R.H., Barth, P.T. & Collins, J. (1969). Control of DNA synthesis in bacteria. Symposium of the Society for General Microbiology, 19, 263-97.
Rollins, M.J., Harper, J.D.I. & John, P.C.L. (1983). Synthesis of individual proteins, including tubulins and chloroplast membrane proteins, in synchronous cultures of the eukaryote Chlamydomonas reinhardii. Journal of General Microbiology, 129, 1899-919.
Sachs, J. (1887). Lectures on the Physiology of Plants. Oxford: Clarendon Press.
Sato, Ch. (1976). A conditional cell division mutant of Chlamydomonas reinhardii having an increased level of colchicine resistance. Experimental Cell Research, 101, 251-9.
Schel, J.H.N., Steenbergen, L.C.A., Bekers, A.G.M. & Wanka, F. (1978). Change of the nuclear pore frequency during the nuclear cycle of Physarum polycephalum. Journal of Cell Science, 34, 225-32.
Schekman, R. (1982). The secretory pathway in yeast. Trends in Biochemical Sciences, 7, 243-6.
Smith, J.A. & Martin, L. (1973). Do cells cycle? Proceedings of the National Academy of Sciences, USA, 70, 1263-7.
Sompayrac, L. & Maaløe, O. (1973). Autorepressor model for control of DNA replication. Nature, New Biology, 241, 133-5.
Spudich, J.L. & Sager, R. (1980). Regulation of the Chlamydomonas cell cycle by light and dark. Journal of Cell Biology, 85, 136-45.
Swift, H. (1953). Quantitative aspects of nuclear nucleoproteins. International Review of Cytology, 2, 1-76.
Tamiya, H. (1963). Cell differentiation in Chlorella. Symposia of the Society for Experimental Biology, 17, 188-214.
Tovey, M. & Brouty-Boye, D. (1976). Characteristics of the chemostat culture of murine leukemia L1210 cells. Experimental Cell Research, 101, 346-54.
Van't Hof, J. & Kovacs, C.J. (1972). Mitotic cycle regulation in the meristems of cultured roots: the principal control point hypothesis. Advances in Experimental Medicine and Biology, 18, 15-30.
Warr, J.R. & Quinn, D. (1977). Low molecular weight sulphydryl components and the expression of a cell division mutant of Chlamydomonas reinhardii. Experimental Cell Research, 104, 442-5.
Wheals, A.E. (1980). Sloppy size control of the Saccharomyces cerevisiae cell cycle. Society for General Microbiology Quarterly, 8, 28.
Woodward, J., Rasch, E. & Swift, H. (1961). Nucleic acid and protein metabolism during the mitotic cycle of Vicia faba. Journal of biophysical and biochemical cytology, 9, 445-62.
Yen, A. & Pardee, A.B. (1979). Role of Nuclear Size in Cell Growth Initiation. Science, 204, 1315-7.
Yeoman, M.M. (1976) (ed.) Cell Division in Higher Plants. London & New York: Academic Press.

THE CONTROL OF THE CELL CYCLE IN RELATION TO FLORAL INDUCTION

D. Francis & R.F. Lyndon

INTRODUCTION

Following the arrival of the floral stimulus, a number of events occur in the shoot apex which commit it to flowering (Bernier, Kinet & Sachs, 1981) and are defined as evocation (Evans, 1971). An increase in the growth rate and changes in the mitotic index are two events which have been consistently found in evocation. Since cell size in apices seems to remain more or less constant, the changes in growth rate imply changes in the mean length of the cell cycle. Changes in mitotic index also indicate alterations in the component phases of the cell cycle. The questions posed here are therefore (1) how does the cell cycle in the shoot meristem change during floral evocation? (2) what are the regulating agents and how is the cell cycle regulated? and (3) what is the significance of these changes to flowering?

THE CELL CYCLE IN VEGETATIVE AND FLOWERING APICES

The vegetative apex of most higher plants typically has a zonate structure. The central zone at the summit of the meristem stains weakly with RNA- and protein-specific dyes, reflecting a lower concentration of RNA and protein relative to the peripheral zone on the flanks of the apex where leaves are initiated. As cells are displaced downwards, by growth, from the central to the peripheral zone their cell cycles shorten (Lyndon, 1973, 1976). In *Chrysanthemum segetum* and *Rudbeckia bicolor* G1 is shorter in the peripheral zone although in *Rudbeckia* G2 is also shortened but not to the same extent. In *Pisum sativum*, G1, S and G2 are all shortened in the peripheral zone although mitosis itself is of the same duration in both central and peripheral zones, as in the other two species (Table 1).

On flowering, an increase in the rate of growth and cell division typically occurs just before flower initiation in a range of unrelated species; this implies a shortening of the cell cycle (Lyndon & Francis, 1985).

Alterations in the cell cycle on flowering are also indicated by transient increases in the mitotic index which are some of the first events in shoot apices during evocation (Bernier, 1971). The first mitotic "burst" may occur very soon after the floral stimulus is presumed to reach the apex, and characteristically there is a second mitotic "burst" about 20-30 h later. An increase in labelling index (the percentage of cells labelled with DNA precursors) after the first mitotic peak suggests that cells may be progressing through a synchronous cell cycle from the first to the second mitosis. However, only for *Sinapis alba* and another LD plant *Silene coeli-rosa*, is there detailed evidence of synchronous divisions and synchronous cell cycles during evocation.

SYNCHRONISATION OF CELL DIVISION

Cell cycle in evocation in Sinapis

The mean length of the cell cycle in vegetative *Sinapis* plants is 288 h in the central zone and 157 h in the peripheral zone of the shoot apex (Bodson, 1975). It reduces to 35 h and 25 h respectively, when the first flower primordia are just being initiated. The shortening of the cell cycle is first observed as a wave of mitosis about 26 h after the beginning of induction (Fig. 1), without a prior wave of DNA synthesis, implying that cells already in G2 come into division, presumably because of a shortening of G2 (Kinet, Bernier & Bronchart, 1967; Bernier, Kinet & Bronchart, 1967). Cells accumulate in G1 (Jacqmard & Miksche, 1971), then there is a peak of

Table 1. Cell cycles in the central zone (CZ) and peripheral zone (PZ) of vegetative shoot apices

		Cell cycle phases (h)					
		G1	S	G2	M	C	
Chrysanthemum segetum	CZ	115	10	7	3	135	Nougarède & Rembur (1978)
	PZ	32	8	8	3	51	
Pisum sativum	CZ	38	13	17	1	69	Lyndon (1973)
	PZ	15	8	4	1	28	
Rudbeckia bicolor	CZ	>40	19	14	<1	>>40	Jacqmard (1970)
	PZ	9	12	9	<1	30	

Fig. 1. Mitotic index (%), labelling index (%) and G2/G1 (>3C:<3C amounts of DNA) ratio in the peripheral zone of the shoot apex of 65-day old plants of <u>Sinapis alba</u> exposed to (a) 1LD and 2SD (▲;●) or (b) SD (Δ), (adapted from Bernier, Kinet & Bronchart, 1967; Jacqmard & Miksche, 1971).

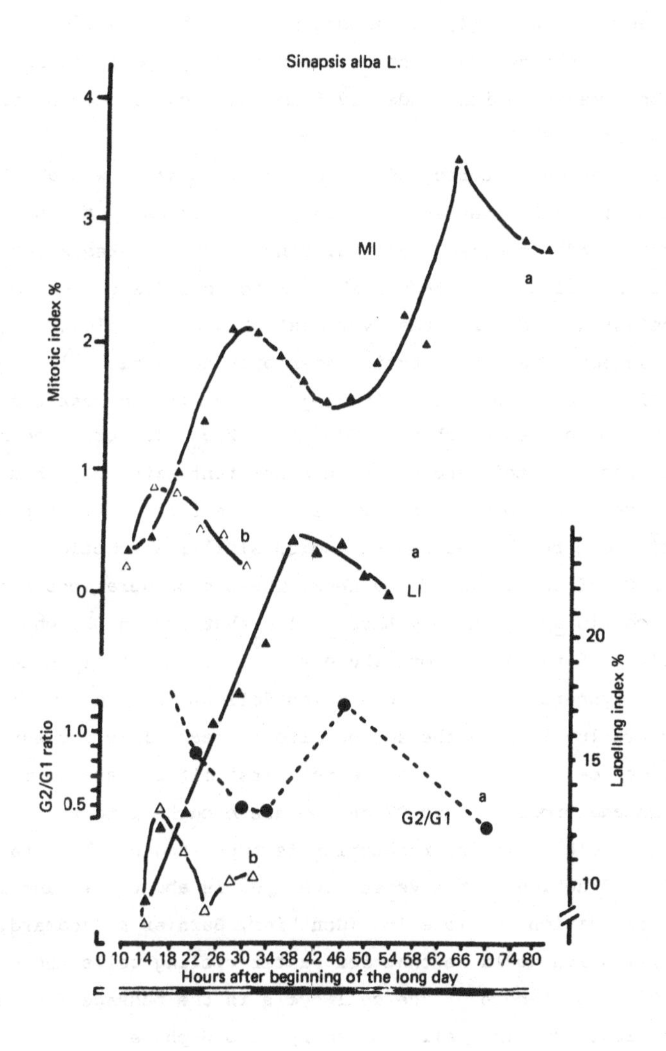

labelling index indicating a wave of cells in the DNA synthetic (S)-phase of the cell cycle (Bernier et al., 1967). This is followed by a preponderance of cells in G2, and then the second mitotic peak about 34 h after the first. These events are consistent with a synchronous cell cycle during the interval between the mitotic peaks, corresponding with the values of 35 h for the mean cell cycle of the central zone in the flowering apex, and 25 h in the peripheral zone, measured independently from the accumulation of colchicine-metaphases (Bodson, 1975).

The maximum mitotic index measured in Sinapis was about 4% (in the second mitotic peak) whereas in a truly synchronous system one would expect a mitotic index approaching 100%. Since mitosis takes about one hour (Bodson, 1975) and since the first peak of mitotic index of just over 2% appears to be normally distributed over most of a cell cycle (Fig. 1), it would be equivalent to a mean mitotic index over the whole cell cycle of about 1.3%. If all cells were contributing to the mitotic peak the mean mitotic index should then be about 100/35 i.e. 2.9%. The observed mean value of 1.3% falls short of this and would be consistent with only about 45% of the cells of the shoot apex being involved in the first mitotic peak. The second mitotic peak reaches almost 4% and on similar assumptions could account for 100% of the cells of the apex. These considerations are consistent with the conclusion that only those cells that are in G2, when the floral stimulus arrives at the apex, contribute to the first mitotic peak (and perhaps to the synchronous cell cycle which follows) but that most or all of the cells are involved in the second mitotic peak. This suggests that the remaining 55% of cells not involved in the first mitosis are stimulated to synthesise DNA and progress into G2 and so are brought into the shorter (25-35 h) cell cycle. This interpretation is supported by the data showing that the growth fraction in the vegetative apex is about 34%, and increases to 61% after transition to flowering (Gonthier, Bernier & Jacqmard, 1984). A further implication would be that for faster cycling cells the control point regulating the length of the cell cycle is the passage from G2 to mitosis, and for slower cycling cells the entry into S-phase.

The relationship between the occurrence of successive peaks of mitosis and subsequent flowering has been examined in a series of experiments in Sinapis (Bernier, et al., 1974). In all the experiments attempted it was impossible to dissociate the flowering process from the occurrence of the first mitotic peak which therefore may be essential for evocation in Sinapis. However, the first mitotic peak could occur without necessarily being followed by flowering, since plants given a 12 h LD, showed the first mitotic peak although not the second, and did not flower. In all plants which subsequently flowered there was a second mitotic peak and this occurred simultaneously with the initiation of flower buds. Synchronisation of the cell cycle may therefore be essential for flowering in *Sinapis*.

Cell cycle in *Silene*

Silene is a qualitative LD plant in which nine days, and 10-11 cell cycles elapse between the beginning of induction and the appearance of sepals (Miller & Lyndon, 1976). Measurements of the rates of growth of the apex suggested that there was a transient increase in the rate of cell division during the first LD and a more marked increase on about the eighth day after the start of induction (Table 2). Synchronisation of cell division was found to occur on day eight, just before the appearance of sepals on day nine. Approximately 20-50% of the cells became synchronised. Between 1500 h of day eight and 0100 h of day nine successive peaks were recorded for the mitotic index, proportion of cells in G1, DNA synthesis (labelling index), proportion of cells in G2, and mitotic index once more (Francis & Lyndon, 1979). The 10 h interval between successive mitotic peaks was identical to the length of the cell cycle measured for the cells of the apical dome by an

Table 2. Changes in cell cycle length in the *Silene* apex

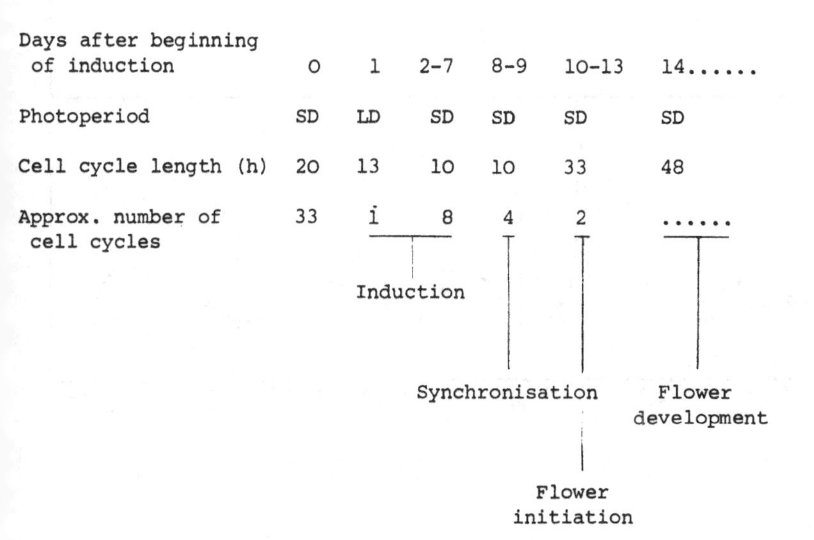

Days after beginning of induction	0	1	2-7	8-9	10-13	14......
Photoperiod	SD	LD	SD	SD	SD	SD
Cell cycle length (h)	20	13	10	10	33	48
Approx. number of cell cycles	33	1	8	4	2	......

independent method (Miller & Lyndon, 1975). This suggests that in *Silene*, all the cells had cycles of a similar length but that, as in *Sinapis*, only a proportion of the cells were synchronised at the time of the first mitotic peak. In *Silene*, later work has shown that there are apparently two synchronous cell cycles, after which synchrony begins to disappear (Fig. 2).

In *Silene*, unlike *Sinapis*, synchrony of division and of cell cycles can be dissociated from flowering. When *Silene* plants were given 7 LD followed by 48 h darkness the mitotic index fell almost to zero, the apices temporarily stopped growing and synchrony did not occur (Fig. 2). When the plants were then placed in SD, growth was resumed (without synchronisation

Fig. 2. Mitotic index (%) and G2/G1 ratio in the apical dome of 28-day old *Silene* plants exposed to 7 LD and 5 SD (days 8-12) (closed symbols-solid line) or exposed to 7 LD, 48h darkness (days 8-9), and then 3 SD (open symbols-dotted line) (days 10-12). 50% of the 7 LD plants had 3 sepals by day 10 (↓) whilst 50% of the dark treated plants had 3 sepals by day 12 (⇩). (Adapted from Grose & Lyndon, 1984).

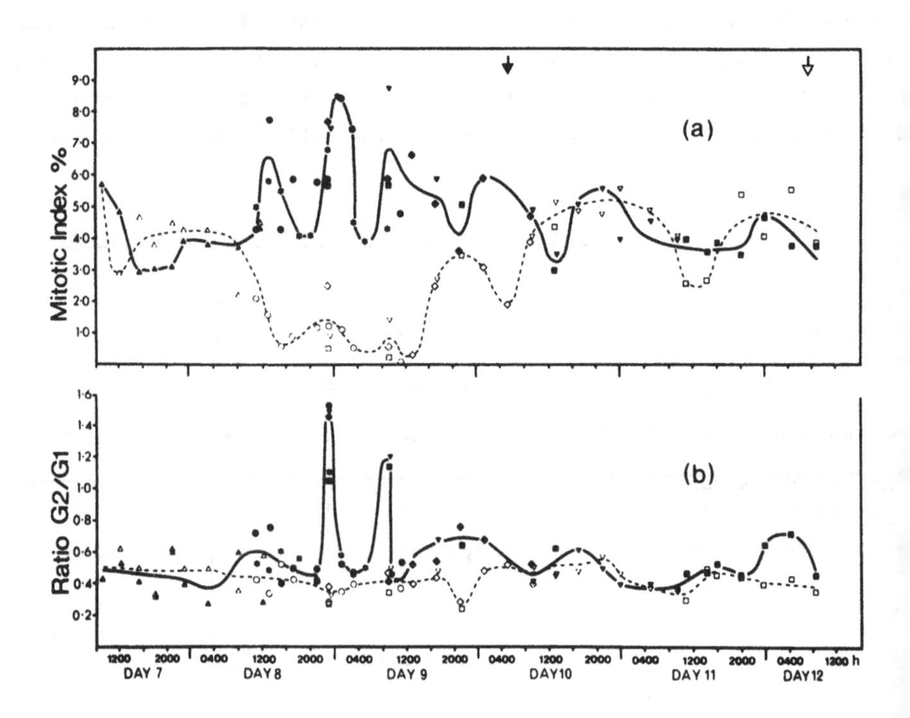

of cell division) and flowers were initiated but with a delay of two days (Grose & Lyndon, 1984). The delay of flowering suggests that some growth had to take place subsequent to the end of induction for flowers to form, perhaps to allow the synthesis of molecules necessary for flowering but which are made only when the cells are actively going through the cell cycle.

The shorter cell cycle in Silene is detectable as a marked increase in cell number which has already occurred by the beginning of day eight (Miller & Lyndon, 1976) whereas the first synchronous cell cycle does not begin until late on day eight (Fig. 2) (Francis & Lyndon, 1979; Grose & Lyndon, 1984). This suggests that synchronisation may be imposed on a set of cells which are rapidly dividing.

Synchronisation in other species

In Xanthium (Jacqmard, Raju, Kinet & Bernier, 1976), Anagallis (Taillandiér, 1978) and other plants, in the absence of information about the length of the cell cycle, it is impossible to be sure that successive mitotic peaks represent anything other than, at most, a partial synchronisation of small numbers of cells without a significant degree of synchrony. Until more information is forthcoming it may be incautious to assume that synchrony of cell division and the cell cycle is always characteristic of floral evocation except on a limited scale in each apex.

Evoked apices in which the cells become synchronised are unusual in that synchrony seems to be the result of, or superimposed on, cells which are growing faster and have shortened cell cycles. This unusual feature could perhaps be because only at evocation does the flow of substances reaching the apex change relatively suddenly, i.e. when components of the floral stimulus arrive.

Even if general synchronisation of cells throughout the apical meristem during evocation is the exception rather than the rule, the occurrence of mitotic peaks during evocation may be consistent with the occurrence of clusters of mitoses in particular regions of the apical dome. Such subpopulations of cells could possibly be concerned with the future initiation of floral organs, involving alterations in the polarity of growth which may relate to changes in the arrangement of primordia which occur at the beginning of flower morphogenesis (Lyndon, 1978).

CELL CYCLE DURING EARLY EVOCATION

In Sinapis the synchronous cell cycle occupies the whole of

evocation and extends up to the initiation of flower buds. Since the "point of no return" of commitment of the apex to flowering is about 44 h after the start of induction (Kinet, Bodson, Alvinia & Bernier, 1971) and just after the peak of DNA synthesis in the synchronous cell cycle, only the events of the cell cycle occurring before the peak of DNA synthesis can be concerned with the processes of evocation.

In *Silene*, in addition to an increase in growth rate on the eighth day after the start of induction, the growth rate was also faster during the first LD (Miller & Lyndon, 1976; Francis & Lyndon, 1978a), but this was not associated with synchronisation of cell division. In the plants exposed to 1 LD, G2 was longer than in the SD controls but G1 had become shortened from 10.0 to 3.8 h (Fig. 3) and so cells accumulated in G2 for about 12-14 h (J.C. Ormrod & D. Francis, unpublished data). The initial accumulation of cells in G2 could be detected as early as one hour after the beginning of the photoextension (Francis, 1981a). Since this consisted of 16 h low intensity light from tungsten bulbs and was preceded by 8 h high intensity light from fluorescent tubes and tungsten bulbs, there was a

Fig. 3. Component phases of the cell cycle (h) in cells of the apical dome of 28-day old *Silene* plants exposed to (a) 1 LD or (b) exposed to 5 min far-red light at 1700h (FR) or (c) maintained in SD (adapted from Ormrod & Francis, 1985; J.C. Ormrod & D. Francis unpublished).

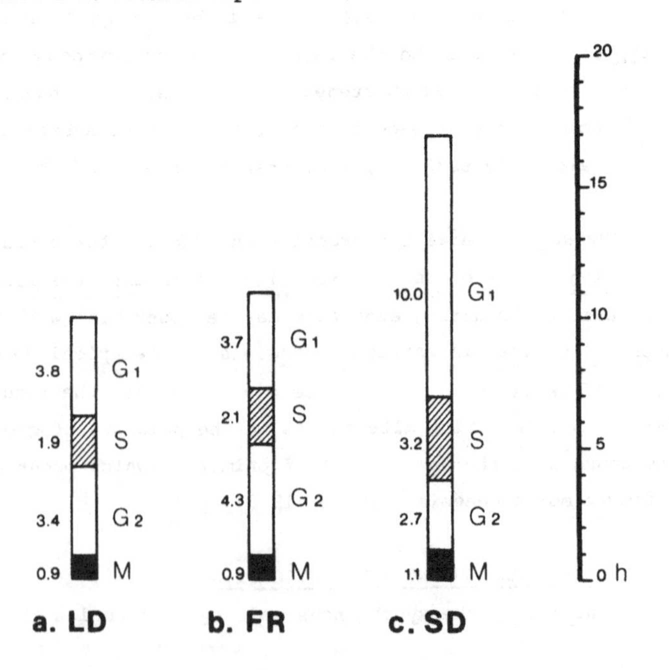

change in spectral quality which involved a decrease in the red/far red ratio (Smith, 1975; J.C. Ormrod & D. Francis, unpublished data). _Silene_ was subsequently found to respond to a 5 min exposure to far-red light by an increase in the proportion of cells in G2 in the apex 3 h later. A short exposure to far-red light therefore effectively replaced the 16 h photo-extension in this respect (Francis, 1981b). Moreover, a similar exposure to red light or far-red followed by red light did not result in a G2 increase and the data were consistent with a low irradiance phytochrome response.

Interestingly, all of the various red and far-red exposures resulted in significant increases in the mitotic index relative to the untreated controls and so are not simple phytochrome effects in which far-red negates an effect of red light. Similar observations have been made on vegetative shoot meristems of etiolated _Oryza sativa_, in which the rate of cell division was stimulated by both red and far-red light (Rolinson & Vince-Prue, 1976). Since the mitotic index in the shoot meristem in _Silene_ is proportional to the rate of cell division, increases in the mitotic index in response to both red and far-red light were consistent with a shortening of the cell cycle relative to the untreated control plants. However, in the two different treatments the shorter cell cycle would be achieved in different ways since far-red light resulted in an early accumulation of cells in G2 but red light did not (Francis, 1981b). Measurements of the duration of the cell cycle and its component phases in _Silene_ have shown that both far-red, and red, light result in a shortened cell cycle, but with a marked shortening of G1 in response to far-red light (Fig. 3). In response to red, there is a dramatic shortening of G2, at least for a sub-population of cells within the apex (Ormrod & Francis, 1985). The shortened cell cycle brought about by far-red light reverts to a longer cell cycle comparable to, and at the same time as, that which occurs in untreated plants given 1 LD (Francis & Lyndon, 1978b). The effect of 1 LD on the cell cycle can therefore be ascribed to the effects of far-red light in the first 5 min of this LD.

The increase in the proportion of cells in G2 only 2 h after a 5 min exposure to far-red light poses the question of how this is achieved by the cells. There is evidence that S-phase of the cycle is shorter when the cell cycle is shortened during the first LD (Francis & Lyndon, 1978b; Ormrod & Francis, 1985). In 2 h the cells therefore must leave G1 faster than they enter it and rapidly traverse S-phase. This may be achieved by faster rates of DNA replication or activation of shorter replicons (see Francis, Kidd & Bennett, this volume). The reduction in the length of G1 could be either because the cells complete the necessary G1 events more rapidly in response to far-red light, or because on leaving mitosis the nuclei are already able to replicate their DNA, and simply do so almost immediately when far-red light has been received. There are examples from both animal and plant systems where G1 is bypassed or omitted completely from the cell cycle showing that G1 is not always a prerequisite for DNA replication (Clowes, 1967; Prescott, Liskay & Stancel, 1982).

Since one LD and the far-red treatment are both non-inductive in *Silene* (Miller & Lyndon, 1976; Francis, 1981a), what, if any, is the relationship between the rapid phytochrome-mediated accumulation of cells in G2 during the first LD (day 0) and the appearance of flower primordia 8 days and 9-10 cell cycles later? Similar accumulations of nuclei in G2 occur at 2000 h on the first 3 of 7 LD (i.e. days 0-2) but from days 3-6 these accumulations can no longer be detected in the apical dome. However, at 2000 h of days 7 and 8, a G2 increase reappeared in the apex even if the LD treatment was ended after 7 LD (J.C. Ormrod & D. Francis, unpublished data). These G2 increases occurred at exactly the time expected for cells synchronised after a normal induction of 7 LD (Francis & Lyndon, 1979), suggesting that the occurrence of synchrony of division on days 8 and 9 is linked to the effect of far-red light during the first 5 min of the photo-extension on each day of induction.

The importance for flowering, of the events triggered by far-red light in the first few minutes of the first 3 days of induction in *Silene*, is confirmed by experiments in which the plants were exposed to short periods of darkness during an otherwise inductive treatment. When *Silene* was given 7 LD but with a 20 min dark period at the beginning of the 16 h photoextension each day, flowering was suppressed (Taylor, 1975) or reduced to 8% of that of the controls in a normal 7 LD regime (J.C. Ormrod & D. Francis, unpublished data). Giving the dark period on only the first 3 of the 7 LD resulted in only 10% flowering and samples taken at 2000 h on each of these dark-interrupted LD showed that a G2 increase did not occur (J.C. Ormrod & D. Francis, unpublished data). Thus, the G2 increases occurring during 3 normal LD were correlated with subsequent flowering of 90-92% of the plants. These results are consistent with the hypothesis that in the apical dome of *Silene* during the first 3 of 7 LD, essential events are occurring which commit the vast majority of the plants to flower, and that these events are correlated with, and perhaps dependent upon, changes in the cell cycle.

Although a G2 increase was not detected during these dark-interrupted long days, the mitotic index in the apical dome increased significantly, consistent with a shortened cell cycle relative to the SD controls. A major difference between this shortened cell cycle and that occurring during 1 LD seems to be that whilst G2 was 34% of the LD cycle, it was only about 27% of the cell cycle in plants given the dark-interrupted LD (J.C. Ormrod & D. Francis, unpublished data). This suggests that in *Silene* it is the accumulation of nuclei in G2 which is correlated with flowering rather than the shortened cell cycle *per se*. Perhaps this protracted time spent in G2 (Fig. 3) is when a qualitative change in gene expression takes place in the apex of *Silene*.

In order to activate DNA synthesis, chromatin may have to undergo conformational changes (Bryant, 1982; Kelly & Trewavas, this volume), the most obvious being the dissociation of DNA from the histones and the subsequent separation of the strands of the DNA molecule. Changes in chromatin structure may therefore modulate the events of DNA replication (De Pamphilis & Wassarman, 1980). Changes in chromatin structure have indeed been found during evocation; in *Sinapis*, the amount of dispersed chromatin reached maxima just before the peaks of mitosis (Havelange & Bernier, 1974). An increase in the dispersion of the chromatin occurred first in G1 nuclei 30 h after the beginning of induction, at about the time of the first mitotic peak, and then later also in the G2 nuclei (Havelange & Jeanny, 1984). Phytochrome may therefore modulate reactions in the nucleus which result in a rapid decondensation and modification of chromatin. It would be of obvious interest to know whether in *Silene* there are also changes in the degree of condensation which result from exposure to far-red light and are associated with the changes in the cell cycle during the first few LD. However, it is still not at all clear what specific changes in gene expression would be expected during evocation (Lyndon & Francis, 1985). Changes in chromatin structure are likely, in any case to result in a general activation or de-activation of many genes (Kelly & Trewavas, this volume). If changes in phytochrome are also affecting ionic fluxes or changes in the concentrations of growth substances in the cells of the apex, then changes in the cell cycle could perhaps be indicators of other underlying changes in the cell rather than themselves being the important events of evocation.

EFFECTS OF PLANT GROWTH REGULATORS ON THE CELL CYCLE DURING EVOCATION

Plant growth regulators have a wide range of effects on flowering ranging from the promotion of flowering by gibberellins in some LD plants and some trees, to complete inhibition. When inhibition of flowering involves inhibition of growth of the shoot apex then naturally the cell cycle in the apex will be affected in some way. However, there are relatively few instances where plant growth regulators have been shown to affect the cell cycle in the shoot apex in a way that is not simply part of a general inhibition of growth. Axillary buds inhibited from growing by apical dominance are usually arrested in G1 (e.g. Cottignies, 1979) and to the extent that auxin or abscisic acid may be responsible for the arrest, these growth substances can be regarded as inhibiting the onset of DNA synthesis (see also Dunham & Bryant, this volume). In apices which have the potential to respond to an inductive stimulus only the effects of gibberellic acid and cytokinins on the cell cycle in the shoot apex have been studied.

Gibberellic Acid

In *Rudbeckia* the application of gibberellic acid causes the plant to bolt and flower and also causes increases in the mitotic index and the labelling index in the central zone and subapical pith of the shoot apical meristem (Bernier, Bronchart & Jacqmard, 1964; Jacqmard, 1968). The increase in labelling index seemed to precede the increase in mitotic index by several hours suggesting that the primary action of GA_3 was to induce the release of nuclei from G1 to S. Also, because the peak of DNA synthesis was only about 2 h before the peak of mitosis in the central zone and subapical pith, GA_3 probably resulted in G2 shortening, thereby releasing nuclei from G2 into mitosis. The parallel effect of gibberellic acid on the cell cycle in the shoot apex, and on promoting flowering, are probably linked. Gibberellic acid increased the rate of cell division in the central zone (as indicated by the increase in mitotic index) which is characteristically activated on flowering in the Compositae, of which *Rudbeckia* is a member (Nougarède, 1967).

Cytokinins

In plants in which application of cytokinin can promote flowering, nothing is known about the effects of cytokinin in the shoot apex. Conversely, in the plants in which the effects of cytokinin on the cell cycle in the meristem have been studied, cytokinin is ineffective in promoting flowering. In *Sinapis*, benzylaminopurine or zeatin applied to vegetative apices resulted in an increase in the mitotic index 10 h later, but without a subsequent increase in DNA synthesis. Moreover, the cytokinin-treated plants did not flower (Bernier, Kinet, Jacqmard, Havelange & Bodson, 1977). Because the cytokinin mimicked the increase in mitotic index which occurred in induced plants, it was suggested that a cytokinin may be the component of the floral stimulus responsible for the initial synchronisation of cell division during evocation.

The effect of cytokinin in the shoot apex is clearly to promote the progression of cells from G2 to mitosis, as it does generally in other tissues and other organisms (Fosket, 1977). Whether cytokinin is part of the floral stimulus or whether it can merely substitute for a component of the floral stimulus is not clear from these experiments.

CONTROL OF PROGRESSION THROUGH THE CELL CYCLE IN RELATION TO FLOWERING

A characteristic feature of evocation is a faster rate of growth of the shoot meristem, often reflected in an increase in the mitotic index. Because in any particular species the duration of mitosis remains constant, the mitotic index is a measure of the rate of cell division and is inversely proportional to cell cycle duration. The implication is that increases in the mitotic index are accompanied by decreases in the lengths of the component phases of the cell cycle. However if it is an increase in growth rate of the apex just before flowering which is important, perhaps in order to allow a new pattern of primordial arrangement to be set up, then the change in the lengths of the phases of the cell cycle may be a general response to a speeding up of growth. There could perhaps be no specific control points in these cycles; they could be regulated by the general rate of metabolism. Similarly although the imposition of synchrony in the shoot meristem of *Silene* implies control of some of the cell cycles in order to get them in phase, these controls may be inessential for flowering (Grose & Lyndon, 1984; Fig. 2).

In both *Silene* and *Sinapis*, synchronisation of cell division is at its most complete immediately before floral morphogenesis. By this time, an entire sequence of events has already taken place to commit the apex to make a flower. The most probable interpretation seems to be that the arrival of substances at the apex (perhaps cytokinins or gibberellins), which are necessary for flowering to occur, also have as a side effect the control of the passage of cells from one phase of the cell cycle to another, so that synchrony results fortuitously, at least in _Silene_.

In _Silene_, the changes that occur in the component phases of the cell cycle during the first 3 of 7 LD are not so easily reconciled to a simple requirement for faster rates of growth since it is not the shorter cell cycle which is correlated with subsequent flowering, but the accumulation of cells in G2. Thus, nuclei accumulate in G2 in response to LD not merely as a consequence of a shorter cell cycle but because of different controls on the passage of cells from one phase of the cell cycle to another. The tentative hypothesis is that G2 has a functional role during early evocation in _Silene_. Caution is necessary, however, since it has not yet been possible to show that accumulation or arrest of cells in a particular cell cycle phase leads to a particular developmental event. For example, cells differentiate equally well from G1 or G2 (e.g. Shininger, 1979).

A remarkable aspect of this cell cycle response during early evocation in _Silene_ is the speed with which cells must leave G1 and traverse S-phase so that they can accumulate in G2 whilst G2 nuclei are rapidly entering mitosis. It will be of fundamental interest to elucidate the mechanisms operating during this shortened S-phase which enable rapid DNA replication. Recently, it has been found that nuclei of root-tips of _Pisum sativum_ which permanently arrest in G2 do so because of incomplete DNA ligation (Van't Hof, 1980 and this volume). These permanently arrested G2 nuclei were characterised by DNA with a molecular weight equivalent to replicon size rather than genome size. Thus, the data suggested that these nuclei were arrested in G2 with the vast majority of replicons non-ligated. Clearly, G2 accumulation in an exponentially growing population of cells in a shoot meristem cannot compare with a group of cells in the root which have ceased to grow. However, the suggestion is that ligation of replicons in actively growing cells may be a G2-dependent event.

In _Silene_, S-phase may have to shorten to less than 1 h for cells to rapidly accumulate in G2. This may in turn mean that ligation of replicons for these cells exclusively occurs in G2. Thus, for those cells speeded throug through S-phase, the recovery time for replicon ligation could result in a protracted G2, before these cells divide. Clearly not all cells would be affected in this manner, particularly those which are in late G2 which would be stimulated to divide in response to photo-extension, or a short exposure to far-red light.

Changes in the cell cycle in the apical dome of Silene at the start of floral induction may well be a reflection of other, as yet unidentified, events which are important for floral growth. Nevertheless the rapid accumulation of cells in G2 is at least a useful marker of the competence of the Silene apex to make a flower.

The availability of data for cell cycle behaviour in only three species, Sinapis, Silene and Rudbeckia, clearly highlights a requirement for much more information on a wider range of species. Only then will we be able to discern the relationship between changes in the cell cycle and the transition from vegetative to floral growth. If cell cycle changes do prove to be markers of other important events, then they could be extremely useful in pinpointing when critical changes which result in floral growth are likely to occur in the shoot apex.

ACKNOWLEDGEMENTS

We are grateful to the A.F.R.C. for supporting work through grants AG72/43 to D.F. and AG 15/166 to R.F.L.

REFERENCES

Bernier, G. (1971). Structural and metabolic changes in the shoot apex in transition to flowering. Canadian Journal of Botany, 49, 803-19.

Bernier, G., Bronchart, R. & Jacqmard, A. (1964). Action of gibberellic acid on the mitotic activity of the different zones of the shoot apex of Rudbeckia bicolor and Perilla nankensis. Planta, 61, 236-44.

Bernier, G., Kinet, J.-M. & Bronchart, R. (1967). Cellular events at the meristem during floral induction in Sinapis alba L. Physiologie Végétale, 5, 311-24.

Bernier, G., Kinet, J.-M. & Sachs, R. (1981). The Physiology of Flowering Vol. II. Boca Raton, Florida: C.R.C. Press.

Bernier, G., Kinet, J.-M., Bodson, M., Rouma, Y. & Jacqmard, A. (1974). Experimental studies on the mitotic activity of the shoot apical meristem and its relation to floral evocation and morphogenesis in Sinapis alba. Botanical Gazette, 135, 345-52.

Bernier, G., Kinet, J.-M., Jacqmard, A., Havelange, A. & Bodson, M. (1977). Cytokinin as a possible component of the floral stimulus in Sinapis alba. Plant Physiology, 60, 282-5.

Bodson, M. (1975). Variation in the rate of cell division in the apical meristem of Sinapis alba during transition to flowering. Annals of Botany, 39, 547-54.

Bryant, J.A. (1982). DNA replication and the cell cycle. In Encyclopaedia of Plant Physiology, 14B, ed. B. Parthier, & D. Boulter, pp.75-110. Berlin: Springer-Verlag.

Clowes, F.A.L. (1967). Synthesis of DNA during mitosis. Journal of Experimental Botany, 18, 740-5.

Cottignies, A. (1979). The blockage in the G1 phase of the cell cycle in the shoot apex of ash. Planta, 147, 15-19.

De Pamphilis, M.L. & Wassarman, P.M. (1980). Replication of eukaryotic chromosomes: A close-up of the replication fork. Annual Review of Biochemistry, 49, 627-66.

Evans, L.T. (1971). Flower induction and the florigen concept. Annual Review of Plant Physiology, 22, 365-94.

Fosket, D.E. (1977). The regulation of the cell cycle by cytokinin. In Mechanisms and control of cell division, ed. T.L. Rost & E.M. Gifford, pp. 62-91. Stroudsberg, Pa.: Dowden, Hutchinson & Ross.

Francis, D. (1981a). A rapid accumulation of cells in G2 in the shoot apex of Silene coeli-rosa during the first day of floral induction. Annals of Botany, 48, 391-4.

Francis, D. (1981b). Effects of red and far-red light on cell division in the shoot apex of Silene coeli-rosa L. Protoplasma, 107, 285-99.

Francis, D. & Lyndon, R.F. (1978a). Early effects of floral induction on cell division in the shoot apex of Silene. Planta, 139, 273-9.

Francis, D. & Lyndon, R.F. (1978b). The cell cycle in the shoot apex of Silene during the first day of floral induction. Protoplasma, 96, 81-8.

Francis, D. & Lyndon, R.F. (1979). Synchronisation of cell division in the shoot apex of Silene in relation to flower initiation. Planta, 145, 151-7.

Gonthier, R., Bernier, G. & Jacqmard, A. (1984). Cell proliferation in the meristem of Sinapis during the floral transition. Archives of International Physiology and Biochemistry (in press).

Grose, S. & Lyndon, R.F. (1984). Inhibition of growth and synchronised cell division in relation to flowering in Silene. Planta, 161, 289-94.

Havelange, A. & Bernier, G. (1974). Descriptive and quantitative study of ultrastructural changes in the apical meristem of mustard in transition to flowering. I. The cell and nucleus. Journal of Cell Science, 15, 633-44.

Havelange, A. & Jeanny, J.C. (1984). Changes in density of chromatin in the meristematic cells of Sinapis alba during transition to flowering. Protoplasma (in press).

Jacqmard, A. (1968). Early effects of gibberellic acid on mitotic activity and DNA synthesis in the apical bud of Rudbeckia bicolor. Physiologie Végétale, 6, 409-16.

Jacqmard, A. (1970). Duration of the mitotic cycle in the apical bud of Rudbeckia bicolor. New Phytologist, 69, 269-71.

Jacqmard, A. & Miksche, J. (1971). Cell population and quantitative changes of DNA in the shoot apex of Sinapis alba during floral induction. Botanical Gazette, 137, 364-7.

Jacqmard, A., Raju, M.V.S., Kinet, J.-M. & Bernier, G. (1976). The early action of the floral stimulus on mitotic activity and DNA synthesis in the apical meristem of Xanthium strumarium. American Journal of Botany, 63, 166-74.

Kinet, J.-M., Bernier, G. & Bronchart, R. (1967). Sudden release of the meristematic cells from G2 as a primary effect of flower induction in Sinapis, Natürwissenschaften, 13, 351.

Kinet, J.-M., Bodson, M., Alvinia, A.M. & Bernier, G. (1971). The inhibition of flowering in Sinapis alba after the arrival of the floral stimulus at the meristem. Zeitschrift für Pflanzenphysiologie, 66, 49-63.

Lyndon, R.F. (1973). The cell cycle in the shoot apex. In The cell cycle in development and differentiation, eds. M. Balls & F.S. Billet, pp. 167-83. Cambridge: Cambridge University Press.

Lyndon, R.F. (1976). The shoot apex. In Cell Division in Higher Plants, ed. M.M. Yeoman, pp. 285-314. London and New York: Academic Press.

Lyndon, R.F. (1978). Phyllotaxis and the initiation of primordia during flower development in Silene. Annals of Botany, 42, 1349-60.

Lyndon, R.F. & Francis, D. (1985). The response of the shoot apex to light generated signals from the leaves. In Light and the Flowering Process eds. K.E. Cockshull & D. Vince-Prue (in press). London and New York: Academic Press.

Miller, M.B. & Lyndon, R.F. (1975). The cell cycle in vegetative and floral shoot meristems measured by a double labelling technique. Planta, 126, 37-43.

Miller, M.B. & Lyndon, R.F. (1976). Rates of growth and cell division in the shoot apex of Silene during the transition to flowering. Journal of Experimental Botany, 27, 1142-53.

Nougarède, A. (1967). Experimental cytology of the shoot apical cells during vegetative growth and flowering. International Review of Cytology, 21, 203-351.

Nougarède, A., & Rembur, J. (1978). Variations of the cell cycle phases in the shoot apex of Chrysanthemum segetum L. Zeitschrift für Pflanzenphysiologie, 90, 379-89.

Ormrod, J.C. & Francis, D. (1985). Effects of light on the cell cycle during the first day of floral induction in Silene coeli-rosa L. Protoplasma (in press).

Prescott, D.M., Liskay, R.M. & Stancel, G.M. (1982). The cell life cycle and the G1 period. In Cell Growth. Ed. C. Nicolini, pp. 305-14. New York: Plenum.

Rolinson, A.E. & Vince-Prue, D. (1976). Responses of the rice shoot apex to irradiation with red and far-red light. Planta, 132, 215-20.

Shininger, T.L., (1979). The control of vascular development. Annual Review of Plant Physiology, 30, 313-37.

Smith, H. (1975). Phytochrome and Photomorphogenesis. Maidenhead, U.K., McGraw-Hill.

Taillandiér, J. (1978). La Floraison du Mouron rouge: Arrivée du stimulus floral. Zeitschrift für Pflanzenphysiologie, 87, 395-411.

Taylor, S. (1975). "Factors controlling the flowering of Viscaria candida." Ph.D. Thesis, London University.

Van't Hof, J. (1980). Pea (Pisum sativum) cells arrested in G2 have nascent DNA with breaks between replicons and replication clusters. Experimental Cell Research, 129, 231-37.

THE DNA ENDOREDUPLICATION CYCLES

W. Nagl, J. Pohl & A. Radler

INTRODUCTION

The term "endoreduplication cycle" ("endo-cycle") is used to cover all variants of the mitotic cycle which lead to an increase in DNA (and the degree of ploidy) within the nucleus of a somatic cell (Nagl, 1976a, 1978). In most plant and animal species, differentiation is accompanied by cessation of mitotic activity and cell division, and may be replaced by cell growth and endo-cycles (in a few species, however, DNA elimination cycles can also take place at the diploid or polyploid level; see for instance, Nagl, 1983a).

Endo-cycles occur as various types. The endoreduplication cycle *sensu strictu* is composed of an S period (DNA synthesis or replication period) and a G phase (G means "gap", a period during which no DNA replication takes place). The endomitotic cycle is characterized by structural changes comparable with those seen in mitosis, but the nuclear envelope remains intact throughout the cycle, and no mitotic spindle is formed. The restitution cycle is less strictly controlled than the endoreduplication and the endomitotic cycles. Mitosis begins but is not finished, mainly due to break-down of the mitotic spindle, and a prophasic, metaphasic or anaphasic set of chromosomes decondenses into a G_1 nucleus, whereby the nuclear envelope is irregularly formed around the chromosomes. Restitution nuclei often show bizarre shapes.

Evidently, the genome is very often not completely replicated during an endo-cycle. In other words, we find differential DNA replication: some regions of a chromosome (genes or non-coding, e.g. repetitive DNA fractions, or heterochromatic sections) are less or more often replicated than the remainder. In such a case, polyploidization leads to a DNA amount that is not only quantitatively, but also qualitatively, different from the basic diploid genome.

The ultimate kind of DNA replication within a nucleus without division, is DNA amplification, i.e. the multiple extra synthesis of a gene or of a non-genic sequence. The amplified DNA is either degraded after use (as in the case of rRNA gene amplification in oocytes), or inserted into the chromosome (as, for instance, in the case of amplification of the dihydrofolate reductase gene in methotrexate-resistant mammalian cell cultures).

Fig. 1 summarizes the sequence of the various types of cell cycles, and Fig. 2 shows their results with respect to the genomic state.

Occurrence of endocycles

Endo-cycles and somatic polyploidy in general are widely distributed among organisms of all systematic positions, from ciliates and algae to humans and orchids. Table 1 gives some examples of plants (for a general review see Nagl, 1978). In a few species, all kinds of organs and tissues were checked for endopolyploidy. It was found that nearly all non-meristematic cells have undergone endo-cycles, e.g. 70-80% in *Beta vulgaris*

Fig. 1. The course of the mitotic cell cycle (circle) and of the endocycles as curtailments thereof. G_1, S, G_2 = prereplication, DNA replication and premitotic period of the interphase, P = prophase, M = metaphase, A = anaphase, T = telophase. Am = DNA amplification, EM = endomitotic cycle, ER = endoreduplication cycle, R = restitution cycle, UR = underreplication cycle (Modified from Nagl, 1982).

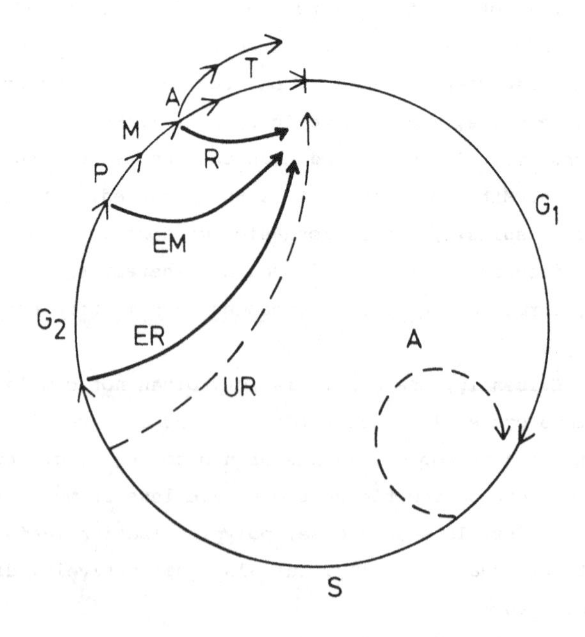

Fig. 2. The results of various types of cell cycles as seen with respect to the genetic state of the resulting cells (only one chromosome per haploid genome is shown) (a) Mitotic cycle: the result is two identical, diploid cells. (b) Mitotic cycle without formation of a cell wall: the result is a binucleate (multinucleate, polyenergid) cell. (c) Endomitotic cycle: the result is a polytene nucleus. (e) Endoreduplication cycle with DNA underreplication: the result is differential polyteny. (f) DNA amplification (extra replication of a gene or other DNA sequence). (g) Somatic reduction (a very rare event, occurring in some insects). (h) Elimination cycle: the result is a reduced chromosome number, or chromatin (DNA) amount, in somatic cells (occurs frequently during cleavage of eggs in insects, but also of other taxa).

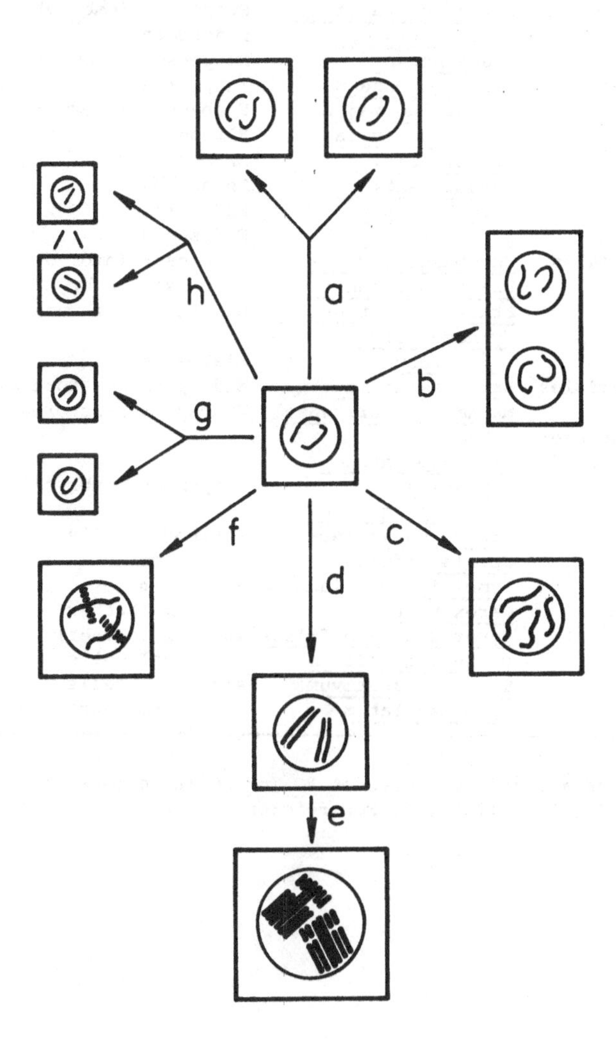

Table 1. Examples for somatic polyploidy in plants (highest levels recorded)

Taxon	Species	Tissue	Degree*
Bryophyta	*Drepanocladus exannulatus*	Paraphyses	8 *C*
Dicotyledoneae:			
Ranunculales	*Aconitum neomontanum*	Antipodal cells	128 *C*
Papaverales	*Papaver rhoeas*	Antipodal cells	128 *C*
	Corydalis cava	Elaiosome	512 *C*
Urticales	*Urtica caudata*	Stinging hairs	256 *C*
Fabales	*Phaseolus coccineus*	Suspensor (embryo)	8,192 *C*
	Phaseolus vulgaris	Suspensor (embryo)	2,048 *C*
	Phlomis viscosa	Endosperm	384 *C*
Myrtales	*Trapa natans*	Suspensor (embryo)	256 *C*
Geraniales	*Geranium phaeum*	Integument	512 *C*
	Tropaeolum majus	Suspensor (embryo)	2,048 *C*
Violales	*Viola declinata*	Elaiosome	16 *C*
Cucurbitales	*Bryonia dioica*	Anther hairs	256 *C*
	Cucumis sativus	Anther hairs	64 *C*
	Cucurbita pepo	Endosperm	284 *C*
	Echinocystis lobata	Endosperm	2,072 *C*
Caryophyllales	*Cucubalus baccifer*	Suspensor (embryo)	128 *C*
	Dianthus sinensis	Suspensor (embryo)	128 *C*
	Echinodorus tenellus	Suspensor (embryo)	128 C
	Gymnocalycium quehlianum	Elaiosome	256 *C*
Scrophulariales	*Bartsia alpina*	Endosperm haustorium	768 *C*
	Melampyrum pratense	Endosperm haustorium	1,536 *C*
Monocotyledoneae:			
Alismatales	*Alisma plantago-aquatica*	Suspensor (embryo)	512 *C*
Liliales	*Allium triquetrum*	Elaiosome	512 *C*
	Crocus suaveolens	Antipodal cells	64 *C*
	Scilla bifolia	Elaiosome	4,096 *C*
Orchidales	*Cymbidium hybridum*	Protocorm	128 *C*
Cyperales	*Carex hirta*	Tapetum (anther)	8 *C*
	Cyperus alternifolius	Tapetum (anther)	8 *C*
Poales	*Hordeum distichum*	Antipodal cells	256 *C*
	Triticum aestivum	Antipodal cells	196 *C*
Arales	*Arum maculatum*	Endosperm haustorium	24,576 *C*

*Values which are not multiples of 2*C* (or 3*C* in endosperm) indicate differential DNA replication. For references see Nagl (1978).

(Butterfass, 1966) and Scilla decidua (Frisch & Nagl, 1979).

The highest levels of polyploidy can normally be found in the ovule, i.e. in synergids, antipodals, endosperm haustoria, and the embryo-suspensor. Fig. 3 shows the embryo-suspensor of Melandrium rubrum with its highly polyploid basal cell.

Whilst the structure of nuclei which have passed through restitution and true endomitotic cycles normally does not differ from that occurring in the diploid ancestor nuclei, the endoreduplication cycle automatically leads to polyteny. In this cycle, the reduplicated chromatids do not separate from each other, because there is no stage during which they condense. (If condensation takes place, for instance in the course of an intercalating mitotic cycle, the polytene chromosomes disintegrate). The structure of plant polytene chromosomes frequently differs from that exhibited by Dipteran salivary gland giant chromosomes, as they do not display a clear banding pattern (for a review see Nagl, 1981). There are two reasons for this. First, some structural changes occur during polytenization in many plant species, so that the chromatids become slightly separated, and the corresponding

Fig. 3. Light micrograph of the embryo-suspensor of Melandrium rubrum. Note the polyploid basal cell. x 500.

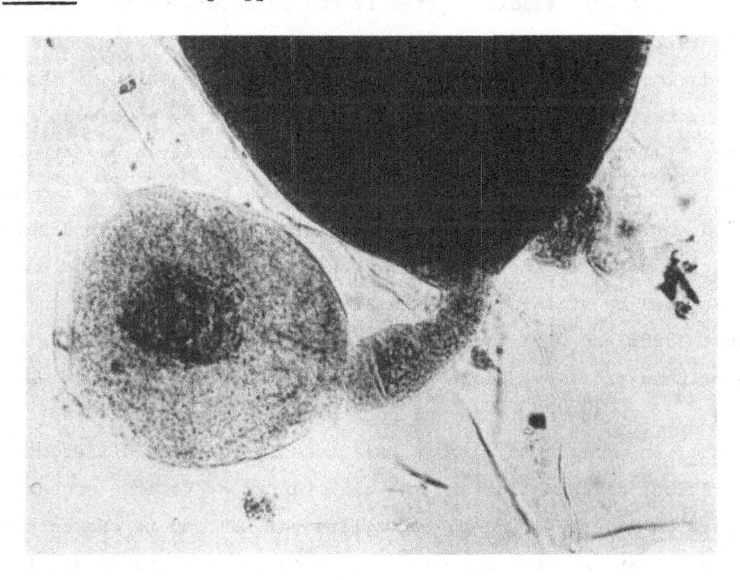

chromomeres are not as exactly adjacent as would be necessary to fuse to a band. Second, the euchromatin of chromonematic nuclei is, in comparison with that in nuclei with different chromatin organization, continuously condensed into the chromonema fiber. As there are no chromomeres and inter-chromomeric regions, no bands and interbands, respectively, can be formed during polytenization. Fig. 4 shows examples of plant polytene chromosomes with and without band formation (in many polytene nuclei, at least the heterochromatic regions of the chromosomes fuse to bands, e.g. in many *Phaseolus* species: Nagl, 1974; in *Ranunculus* species: Tschermak-Woess, 1956; Wedzony, 1982).

DNA UNDER-REPLICATION AND AMPLIFICATION

Specific attention should be drawn to endo-cycles with differential DNA amplification, because they seem to be the basis of many important biological phenomena such as resistance and carcinogenesis, but also "normal" differentiation. Most spectacular are recent results on DNA amplification (e.g. of oncogenes) in the course of carcinogenesis and cell transformation *in vitro* (e.g. Arrighi, 1983). Also, reports of differential DNA replication increase in number every year (Table 3; for a review see Nagl, 1979).

A well studied system is the protocorm of the orchid *Cymbidium* (Nagl, 1972). Here, heterochromatin amplification takes place in certain cell types during certain steps of differentiation. Depending on the hormone supplementation of the culture medium, the mitotic, endoreduplication, or differential DNA replication cycle could be stimulated, and thus different pathways of differentiation could be induced. Fig. 5 shows isolated, Feulgen-stained nuclei from protocorms of *Cymbidium*, some of which display heterochromatin amplification. The DNA of the nuclei differs in AT content as demonstrated by staining with the AT-specific fluorochromes quinacrine, Hoechst 33258 and DAPI (Schweizer & Nagl, 1976), and by analysis of derivative melting profiles (Nagl & Rücker, 1976).

We are now studying both plants grown *in vivo* and cell cultures *in vitro*, in order to find some relationships between differential DNA replication, differentiation, and cell function. Various organs of nasturtium (*Tropaeolum majus*) were used to isolate the DNA and to characterize it by cytological and biochemical methods. It could be shown (Nagl, 1976b, 1978; Nagl, Peschke & van Gyseghem, 1976) that there is a tissue-specific variation of the DNA amount per nucleus and of the DNA composition. Figs. 6 and 7 show respectively the melting profiles and reassociation kinetics of the DNA

Fig. 4. Polytene chromosomes of plants. (a) Polytene chromosome with banded region from the suspensor of Phaseolus coccineus (phase contrast, x 2000). (b) Polytene nuclei in the three antipodal cells of Scilla decidua; here the polytene chromosomes exhibit a cable-like structure (Feulgen, x 230). (Modified from Nagl, 1978).

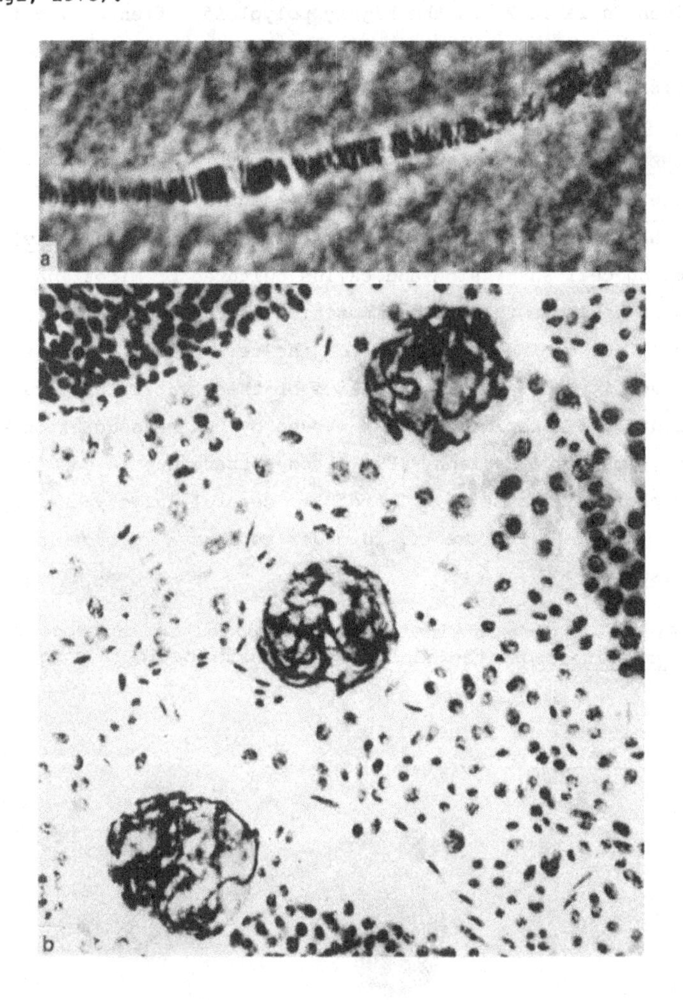

isolated from various organs of Tropaeolum majus. This can only be explained by differential DNA replication, as all possible methodological errors and random variations could be ruled out by employing different methods of DNA isolation and statistical treatment of the results. An analysis of the results is given in Table 2. In the highly polyploid, often polytenic nuclei of the embryo-suspensor, heterochromatin under-replication in certain cells is directly visible at the light microscope level (Fig. 8).

THE ROLE OF PHYTOHORMONES

The role of phytohormones in the control of endo-cycles and differential DNA replication was studied in our laboratory. In in vitro experiments using cell suspension and callus cultures of Nicotiana and Crepis, it could be shown that the composition of the culture medium exerts a significant effect on the DNA content of the cell nuclei. Some of the results are given in Table 3. Although some of these variations are surely due to differential DNA replication, as proven by DNA reassociation kinetics (e.g. Dührssen, Schäfer & Neumann, 1979), some others may be due to changes in nuclear morphology (karyotype). Therefore, careful karyotype analysis must accompany the DNA measurements, in order to find a correct interpretation of the results.

Fig. 5. Feulgen-stained nuclei, isolated from protocorms of Cymbidium. Note the amplified heterochromatin in some nuclei. x 600.

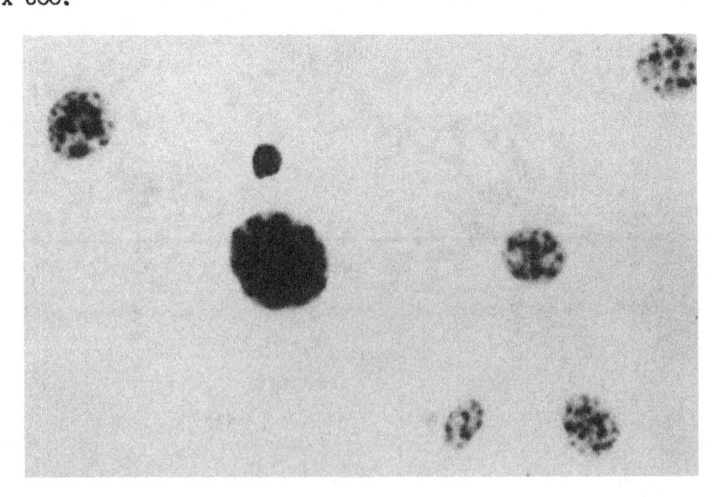

Fig. 6. Derivative melting profiles of DNA isolated from nuclei of various organs of Tropaeolum majus (——leaves, -·-·-fruits, ····stems, ---roots). Mean curves for 15 DNA isolations and denaturations. The differences clearly indicate differential DNA replication.

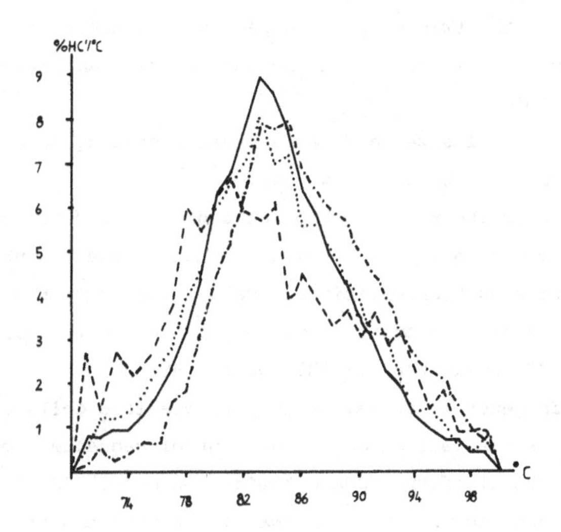

Fig. 7. Reassociation kinetics (cot curves) of the DNA isolated from nuclei of various organs of Tropaeolum majus (from top to bottom: roots, fruits, leaves, stems). Mean curves for 15 DNA isolations and renaturations. The faster the renaturation takes place, the higher is the proportion of repetitive DNA and the frequency of repeats (see also Table 2).

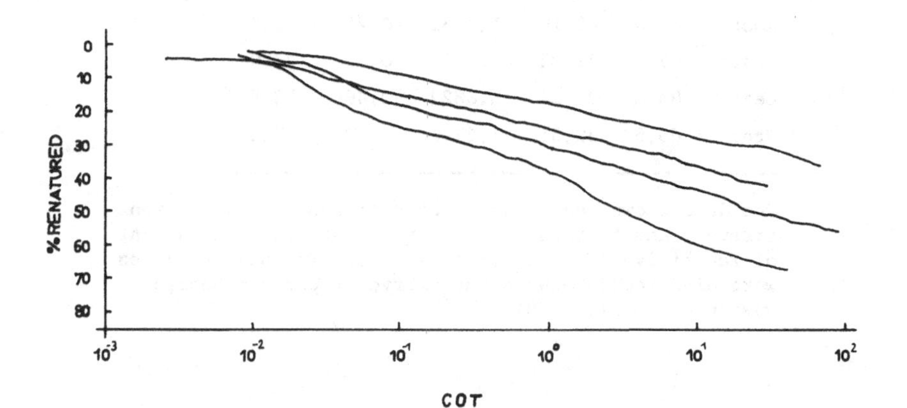

Functional significance of endocycles

The increase in DNA content in somatic nuclei by endo-cycles, with total or differential DNA replication can have various functions. Those which are assumed by most authors to be most significant are the following:
(a) Increase of the transcription capacity, and hence of cell function (this is particularly evident in the case of gene amplification, but also general polyploidy).
(b) Increase of cell size, and hence organ modelling (e.g. in hairs of plant leaves, in scales of butterfly wings).
(c) Alteration of the relationship between genes and non-coding DNA, and hence regulatory changes. It is evident that a huge amount of the non-coding DNA is effective as "conformational DNA", influencing the overall structure of a gene-containing chromatin domain according to thermodynamic laws (Nicolini, 1979; Zuckerkandl, 1981; Nagl, 1983b).

In general, certain highly-specialized cells apparently need huge amounts of DNA (either genes, or control sequences, or both). This fact led to the proposal of the "DNA optimization model" (Nagl, 1977, 1978) that connects the somatic variation in DNA to the evolutionary changes (see also Cavalier-Smith, this volume).

Table 2. Characteristics of the DNA isolated from various organs of the nasturtium, Tropaeolum majus

Organ	Tm °C	GC%	ρ g cm^{-3}	5-mC %	% rep. DNA
Root	86.6	42.19	1.7122	0.26	33.0
Fruit	82.4	31.85	1.6665	0.30	43.0
Leaf	85.0	38.29	1.6622	0.46	52.5
Stem	85.5	39.51	1.6741	0.62	67.0

The data shown that the DNA composition varies between tissues, and that the 5-mC content parallels the amount of repetitive DNA. Differences in the Tm and Cot values were also found between the embryo proper and embryo-suspensor (Nagl, 1978).

Table 3. Examples of effects of phytohormones on the mean nuclear DNA content in plant cell suspension cultures

Species	Hormones	Concentration (mg/l)	DNA Content (in C Values)
Nicotiana tabacum Line TX 1	Control	---	2.00
	IAA	8.0	1.59
	2,4-D	0.5	
	Kinetin	0.1	
	IAA	16.0	2.24
	2,4-D	1.0	
	Kinetin	0.1	
	GA_3	0.2	1.41
	Kinetin	0.1	
Nicotiana tabacum Line TX 4	Control	---	4.59
	IAA	8.0	6.43
	2,4-D	0.5	
	Kinetin	0.1	
	IAA	16.0	6.09
	2,4-D	0.5	
	Kinetin	0.1	
	GA_3	0.2	4.85
	Kinetin	0.1	

Similar results were obtained with *Crepis capillaris* cell cultures. The differences are highly significant according to the Kolmogoroff-Smirnoff test.

The DNA optimization model

The accumulation of data on DNA changes during phylogeny and ontogeny allow a model to be constructed of the relationships between various mechanisms of DNA change during evolution and differentiation. However, it must be emphasized that this model may serve as a guide for further studies rather than as an explanation of the central questions of biology.

The following conclusions can be drawn with some certainty, and therefore be parameters of the model:

1. Evolutionary progress of a biological system cannot be accurately measured except by assessing the complexity of the system. Thus, the more structurally complex an organism becomes, the more advanced it is said to be. However, complexity associated with multicellular, relative to unicellular, organisms is not necessarily accompanied by increases in genic information in the former (see Francis, Kidd & Bennett, this volume). Often the major differences at the DNA level lie in increases in non-genic, repetitive DNA in the more complex organisms; such increases may contribute to evolutionary advances.
2. Adaptation is based upon duplication and mutation of coding genes; the

Fig. 8. Polyploid nuclei in suspensor cells of the nasturtium embryo (*Tropaeolum majus*). While the nucleus on the left exhibits a polytenic structure, the nucleus on the right exhibits only very small chromocenters, indicating DNA (heterochromatin) under-replication. Feulgen, phase contrast, x 1000.

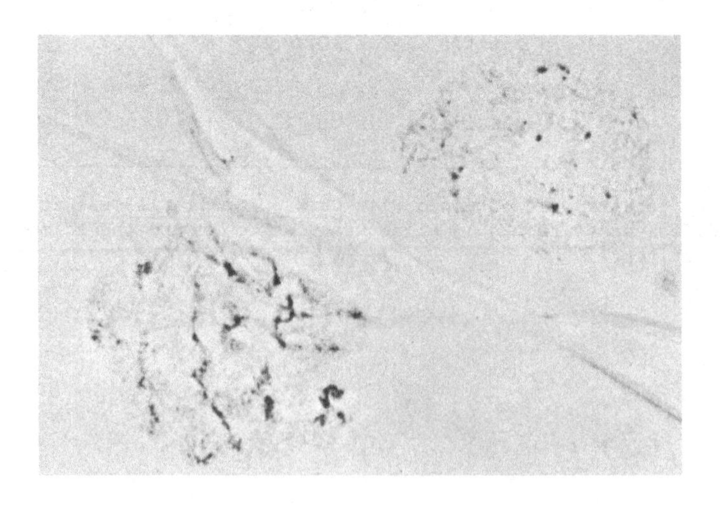

result of this is protein and enzyme polymorphism.

3. Speciation and specialization are characterized by qualitative and quantitative changes, particularly in non-coding, repetitive regulatory sequences. An alternative strategy is karyotype repatterning.

4. The main mechanisms of genome and karyotype evolution are: amplification (saltatory replication) of a non-coding sequence, its diversification and dispersion throughout the genome; heterochromatin polymorphism and centromeric fusions; generative polyploidization and karyotype diversification; addition and loss of B chromosomes; transposition of mobile elements. These mechanisms act as evolutionary strategies, which may be mutually exclusive, or overlap.

5. Highly active cells require a certain minimum DNA content independent of the 2*C* value of a given species. This DNA content can be achieved by somatic polyploidy (including endoreduplication, endomitotic polyploidization, polyteny) and differential DNA replication (DNA amplification, endo-cycles with underreplication).

6. There is a negative relationship between the *C* value and the tendency to polyploidization. This indicates that ontogenetic, lateral increases in DNA can substitute for phylogenetic, tandem increases in DNA.

7. There is selection pressure against mitosis in complex organisms so that the mitotic cycle is gradually reduced through bypass of cytokinesis, some or all mitotic stages, and part of genome replication. Thus the restitution cycle, the endomitotic cycle, the endoreduplication cycle, the underreplication cycle and the amplification cycle evolved in the given sequences.

8. Tissue-specific patterns of endopolyploidy and differential DNA replication are part of the cell differentiation programme. There exist, of course, alternative ways, e.g. at the transcriptional level (activation and inactivation, respectively, of genes), at the post-transcriptional level (e.g. different sites of splicing of hnRNA) and the protein level (e.g. phosphorylation, trimming of proteins). However, differential DNA replication during ontogenesis may be seen as the consequence of a recent phylogenetic change in nuclear DNA, and as the expression of 'the optimal decision in relation to function'.

9. One adaptive value of endopolyploidy (somatic polyploidy) may be seen in the fact that cells of meristems and proliferative tissues can be of low DNA content (which is important for nucleotypic parameters such as fast cell cycles), while highly differentiated and specialized cells often possess many DNA templates due to polyploidization.

10. The evolutionary strategies of speciation and differentiation use the same mechanisms and may be coupled in a feedback regulation system. It seems to be a matter of environmental conditions and genetic background as to whether the mechanism of DNA variation is transitory and somatic, or whether it will be fixed for evolution.

11. There is no DNA constancy, either during evolution or during ontogenetic development. Both the quantitative variation and the qualitative changes of nuclear DNA, and particularly of the non-coding repetitive DNA, may be the key to the understanding of evolution and cell differentiation.

The major point of this model is that it assumes an 'optimum DNA amount' for every cell type of a given organism (see Francis, Kidd & Bennett, this volume). This DNA amount is under selection pressure with respect to a regulatory, transcriptional or nucleotypic advantage. The phylogenetic increase (or decrease) is the result of sequence amplification and divergence, which are primarily selected with respect to their regulatory advantage. The evolutionary programme cannot, however, be viewed independently of the somatic developmental programme, because selection does not act on a species but on its individuals at various developmental stages. Therefore, the DNA data must be processed, in addition, at the ontogenetic level. This is a predefined process, because the decisions depend on the 2*C* value of the species or population. The phylogenetic and ontogenetic programmes are complementary to one another, as the diversity of organs and tissues is the result of phylogenetic diversification of taxa.

Using the terms of Galau *et al*., (1976), one has to assume that divergence during phylogenesis and ontogenesis are two sides of the same coin, or, in other words, speciation and differentiation are two sides of DNA variation. The model proposed is imperfect, but it shows a new way of thinking about somatic DNA values. Experiments should be possible to confirm or refute the DNA optimization model.

The biochemical interactions between ions, hormones, regulatory proteins on the one hand, and the activity of DNA and chromatin with respect to replication and transcription on the other, are only poorly understood. We do not have more than basic descriptions of the events, and therefore cannot yet see the causal relationships between them if any.

Recently, therefore, increasing attention has been drawn to a physical basis of cell cycles and cell differentiation. One of the ideas is given in an accompanying paper (Nagl & Scherthan, this volume). There is accumulating evidence that genotypic changes during evolution and their consequences for differentiation and the phenotype are driven by both "chance" and "necessity" i.e. random mutations, fluctuations and bifurcations, and physical and organismic limitations and constraints (thermodynamics, electrodynamics, system's conditions; see, for instance, Prigogine, 1955; Monod, 1971; Riedl, 1977; Haken 1980; Nagl, 1983c). The concept of selfish DNA (or parasitic or junk DNA) is surely not adequate to explain the high amount of non-coding DNA (up to 99.9%) and of mobile genetic elements in the genome of higher organisms (see also Francis, Kidd & Bennett, this volume). A re-evaluation of the biological significance of DNA variation by endo-cycles in which physical aspects like long-range coherent effects are considered (e.g. Rowlands, 1983) is therefore strongly required.

REFERENCES

Arrighi, F.E. (1983). Gene amplification in human tumor cells. In *Biology of Cancer* Vol. 2, eds. E.A. Mirard, W.B. Hutchinson & E. Mihich, pp. 259-68. New York: Alan R. Liss.

Butterfass, Th. (1966). Neue Aspekte der Polyploidieforschung und Polyploidiezüchtung. *Mittheilungen Max-Planck-Gesellschaft*, I, 47-58.

Dührssen, E., Schäfer, A., Neumann, K.-H. (1979). Qualitative differences in the DNA of some higher plants, and aspects of selective DNA replication during differentiation. *Plant Systematics & Evolution*, Suppl. 2, 95-103.

Frisch, B. & Nagl, W. (1979). Patterns of endopolyploidy and 2C nuclear DNA content (Feulgen) in *Scilla* (Liliaceae). *Plant Systematics & Evolution*, 131, 261-76.

Galau, G.A., Chamberlin, M.E., Hough, B.R., Britten, R.J. & Davidson, E.H. (1976). Evolution of repetitive and non-repetitive DNA. In *Molecular Evolution*, ed. J.F. Ayala, pp. 200-24. Sunderland, Ma. USA: Sinauer.

Haken, H., ed. (1980). *Dynamics of Synergetic Systems*. Berlin: Springer-Verlag.

Monod, J. (1971). *Chance and Necessity*. Munich: Piper.

Nagl, W. (1972). Evidence of DNA amplification in the orchid *Cymbidium in vitro*. *Cytobios*, 5, 145-54.

Nagl, W. (1974). The *Phaseolus* suspensor and its polytene chromosomes. *Zeitschrift für Pflanzenphysiologie*, 73, 1-4'

Nagl, W. (1976a). *Zellkern und Zellzyklen*. Stutgart: Ulmer.

Nagl, W. (1976b). Early embryogenesis in *Tropaeolum majus* L.: evolution of DNA content and polyteny in the suspensor. *Plant Science Letters*, 7, 1-8.

Nagl, W. (1977). The DNA optimization model for speciation and differentiation. *Chromosomes Today* 6, 151-2.

Nagl, W. (1978). *Endopolyploidy and Polyteny in Differentiation and Evolution*. Amsterdam: North-Holland.

Nagl, W. (1979). Differential DNA replication in plants: a critical review. *Zeitschrift für Pflanzenphysiologie*, 95, 283-314.

Nagl, W. (1981). Polytene chromosomes of plants. *International Review of Cytology*, 73, 21-53.

Nagl, W. (1982). DNA endoreduplication and differential DNA replication. In *Encyclopedia of Plant Physiology*, 14B, eds. B. Parthier & D. Boulter pp. 111-24. Berlin: Springer-Verlag.

Nagl, W. (1983a). Heterochromatin elimination in the orchid Dendrobium. Protoplasma, 118, 234-7.

Nagl, W. (1983b). Nuclear organization - physical aspects of gene regulation. In Kew Chromosomes Conference, II. eds. M.D. Bennett & K. Jones, pp. 55-61. London: George, Allen & Unwin.

Nagl, W. (1983c). Evolution: theoretical and physical considerations. Biologische Zentralblatt, 102, 257-69.

Nagl, W. & Rücker, W. (1976). Effects of phytohormones on thermal denaturation profiles of Cymbidium DNA: indication of differential DNA replication. Nucleic Acids Research, 3, 2033-9.

Nagl, W., Peschke, C. & van Gyseghem, R. (1976). Heterochromatin underreplication in Tropaeolum embryogenesis. Natürwissenschaften, 63, 198.

Nicolini, C. (1979). Chromatin structure, from Angstrom to micron levels, and its relationship to mammalian cell proliferation. In Cell Growth, ed. C. Nicolini, pp. 613-66. New York: Plenum.

Prigogine, L. (1955). Thermodynamics of Irreversible Processes. New York: Wiley.

Riedl, R. (1977). A systems-analytical approach to macro-evolutionary phenomena. Quarterly Revue of Biology, 52, 351-70.

Rowlands, S. (1983). Some physics aspects for 21st century biologists. Journal of Biological Physics, 11, 117-22.

Schweizer, D. & Nagl, W. (1976). Heterochromatin diversity in Cymbidium, and its relationship to differential DNA replication. Experimental Cell Research, 98, 411-23.

Tschermak-Woess, E. (1956). Notizen über die Riesenkerne und "Riesenchromosomen" in den Antipoden von Aconitum. Chromosoma, 8, 114-34.

Wedzony, M. (1982). Endopolyploidy and structure of nuclei in the antipodals and syngergids of Ranunculus baudotii Godr. Acta Biologica Cracoviensa Series Botanica, 24, 43-62.

Zuckerkandl, E. (1981). A general function of noncoding polynucleotide sequences. Molecular Biology Reports 7, 149-58.

THE CHLOROPLAST DIVISION CYCLE AND ITS RELATIONSHIP TO THE CELL DIVISION CYCLE

S.A. Boffey

INTRODUCTION

Chloroplast DNA (cpDNA) codes for all of the organelle's rRNAs and tRNAs, and for about 100 of its proteins; it is therefore essential, though not sufficient, for the chloroplast's survival. Since chloroplasts, or plastids, are formed by the division of pre-existing plastids, their DNA must be replicated to ensure the presence of at least one genome copy per plastid. In fact, replication of cpDNA can be so rapid that it will form up to 20% of the cell's DNA, as in spinach leaves (Scott & Possingham, 1983), and up to 1000 genome copies may be present in a single wheat chloroplast (Boffey & Leech, 1982). The minimum number of genome copies per plastid reported in higher plants appears to be about 30 (Table 1). Although it is

Table 1. Levels of plastid DNA in various species

	genome copies/plastid			
	Young	Mature	plastid DNA / total DNA	References
Pea	240	170	8-12%	Lamppa & Bendich, 1979
Spinach	200	30	21%	Scott & Possingham, 1983
Wheat	1000	300	17%	Boffey & Leech, 1982
Beet	100	30	8-11%	Tymms, Scott & Possingham, 1983

not known why plastids contain so many copies of their DNA, the universal occurrence of multiple copies in higher plants suggests that they are needed for normal plant development.

Since the maintenance of adequate levels of cpDNA is clearly essential, it is important to establish if there is a 'chloroplast cycle' in which each round of chloroplast division is accompanied by a round of cpDNA replication, and to determine if these processes are related to the cell cycle. The physical and biochemical mechanisms of plastid division and cpDNA replication must be worked out if we are to understand the regulation of these processes.

CHLOROPLAST DEVELOPMENT IN INTACT TISSUE

The pattern of plastid development has been analysed in several species, with respect to cell division, plastid division, nuclear DNA replication, and chloroplast DNA replication. Monocots, such as wheat, grow from a basal meristem, so that cell age increases with distance from the leaf base; hence chloroplast development can be followed by examining tissue taken from various positions along the leaf. In wheat (Boffey, Ellis, Sellden & Leech, 1979; Boffey & Leech, 1982; Ellis, Jellings & Leech, 1983) it has been been shown that all cell division and nuclear DNA (nDNA) replication occurs in the bottom centimetre of the leaf. Cells are expanding throughout the bottom four cm of leaf, and this expansion is accompanied by a three-fold increase in the number of chloroplasts per cell and at least a 25% increase in the total cpDNA per cell. Thus chloroplasts can divide, and also replicate their DNA, in the absence of cell division or nDNA replication. Similarly, when Rose, Cran & Possingham (1975) incubated discs of spinach leaves with ^{3}H-thymidine, and examined the tissue using autoradiography, they often observed cells in which all the chloroplasts, but not the nucleus, were radioactively labelled.

Although chloroplast and cell divisions do not always occur simultaneously, they must do so in the early stages of leaf development. Wheat and spinach mesophyll cells always contain chloroplasts, and so there must be firstly, a high rate of plastid division and cpDNA replication in dividing cells and secondly, a mechanism to ensure that plastid division at least keeps pace with cell division. It has not been possible to examine this phase of plastid development in wheat, owing to the difficulty of counting very small plastids in the tightly packed dividing cells at the leaf base. However, Scott & Possingham (1983) have investigated these very early stages of development in spinach leaves (Table 2). When the two smallest leaves are compared, it can be seen that cpDNA synthesis occurs at the same rate as nDNA synthesis, resulting in a constant proportion of cpDNA in the cell. There is no increase in the number of plastids per cell, and so the DNA content of each plastid roughly doubles. As leaves expand from 2mm to 20mm, cpDNA replication continues faster than that of nDNA; thus the percentage of cpDNA in the cell increases dramatically. At the same time, chloroplast division takes place, approximately keeping pace with cpDNA replication; and so the number of genome copies per chloroplast rises only slightly, in spite of the huge increase in total cpDNA per cell. In the last phase of development, replication of both nDNA and cpDNA has stopped, but chloroplast division continues, resulting in a diluting out of the cpDNA amongst the increased number of chloroplasts in each cell.

It is therefore clear that chloroplast development is not based on a 'chloroplast cycle', and that replication of cpDNA is not synchronised with that of nDNA. However, there are indications that final cpDNA levels per cell may be related to those of nDNA. The alga *Olisthodiscus luteus* can be induced to produce different numbers of chloroplasts per cell in response to a change in growth conditions. When the levels of nDNA and cpDNA were analysed in such cells, it was found that they remained constant, regardless of chloroplast number, so that the genome copy number per chloroplast was inversely proportional to the number of chloroplasts per cell (Cattolico, 1978). The ratio of cpDNA to nDNA in *Chlamydomonas reinhardii*

Table 2. Development of spinach leaves

Leaf size (mm)	<1	2	20	100
Genome copies per plastid	76	150	190	32
Plastids/cell	10	10	29	171
Genome copies per cell	760	1500	5510	5470
Plastid DNA as % of total	7	8	23	23

(Scott & Possingham, 1983).

was found to be the same in both haploid and diploid strains (Whiteway & Lee, 1977). Algae may not be perfect models for higher plants, but a similar pattern has been observed in etiolated pea seedlings (Lamppa & Bendich, 1979), in which the percentage of cpDNA remained at 1.4%, even when nuclear ploidy increased from 2*C* to 4*C*. When Dean & Leech (1982) measured DNA levels in a non-isogenic polyploid series of wheat, they found the same percentage of cpDNA in the diploid and tetraploid species, even though the number of chloroplasts per cell was twice as high in the tetraploid. However, this relationship was not maintained in the hexaploid, in which the percentage of cpDNA was much lower than that found in the other ploidies. So it seems that the nature of the nuclear DNA, as well as its amount, is important in determining levels of cpDNA per cell. It would be interesting to see the results of a similar experiment performed on an isogenic polyploid series. Unfortunately, such experiments cannot distinguish between a primary effect of nDNA level on cpDNA replication, and secondary effects, since cell area and chloroplast number also seem to increase with ploidy; cpDNA levels could therefore be influenced by cell area or chloroplast number, rather than by nDNA levels.

During the rapid cell division phase of leaf development, some mechanism must exist to ensure adequate rates of chloroplast division and cpDNA replication, compared with the rate of cell division. The relationship between cell and chloroplast division would be most readily studied if cell division were synchronous, and this can be achieved with many algae and lower plants. Rose *et al*. (1975) reviewed such studies, and pointed out that cpDNA synthesis is not restricted to any one phase of the cell cycle. They concluded that there must be independent control of the nuclear and chloroplast DNA polymerases. Direct evidence for this prediction has recently been provided by Aoshima, Nishimura & Iwamura (1983), who have shown that the activities of 'nuclear' and 'chloroplastic' DNA polymerases increase at different times in the cell cycle of *Chlorella*.

USE OF INHIBITORS OR OF INHIBITORY CONDITIONS

One approach to the problem of identifying factors involved in the regulation of chloroplast division and cpDNA replication is to introduce a factor which perturbs some aspect of cellular function, and then see what happens to the developmental processes. The factors which have been used include darkness, inhibitors of protein synthesis, and inhibitors of DNA replication.

Light and dark

When Possingham & Smith (1972) studied the effects of light intensity on chloroplast development in spinach leaf discs, they found that the number of chloroplasts per cell was proportional to the light intensity, up to a saturating level. However, cell area was also proportional to light intensity, so it may be that light stimulates cell expansion and that chloroplasts grow and divide to fill the space available. Possingham (1973a, b, 1980) demonstrated that white light is necessary for chloroplast division in spinach discs, since green light allowed cell growth and an increase in total area of plastids per cell, but no increase in plastid numbers. On transfer to white light, the large plastids produced in green light divided rapidly. The division of pea plastids is also stimulated by light (Bennett & Radcliffe, 1975), but is still able to occur, at a reduced rate, in darkness.

It therefore seems that there may be a relationship between the total chloroplast area and cell size, but the way in which the chloroplast area is 'packaged', whether as large or small chloroplasts, can be influenced by such factors as light. Light has also been shown to stimulate cpDNA replication in peas: in darkness, replication of nDNA and cpDNA proceed at the same rate, but light causes an increase in the percentage of cpDNA in the cell (Lamppa & Bendich, 1979). It has not been demonstrated whether this increase in DNA synthesis occurs in response to light-stimulated chloroplast division, or *vice-versa*; perhaps both processes are regulated independently by light. In this context it is interesting to note that Boasson & Laetsch (1969) reported that fluorodeoxyuridine, which was thought to inhibit cpDNA replication, did not inhibit light-induced chloroplast division in tobacco leaf discs, suggesting that chloroplast division is not triggered by cpDNA replication.

Inhibitors of protein synthesis

The rationale behind the use of inhibitors of protein synthesis is that certain inhibitors, such as chloramphenicol (CAP), will specifically inhibit translation on 70*S* ribosomes, and will therefore selectively block the synthesis of proteins inside the chloroplast. Conversely, cycloheximide (CHI) blocks protein synthesis on the 80S ribosomes of the cytoplasm. Hence, if CAP, but not CHI, blocks a particular step in chloroplast development, that step may depend on the synthesis of one or more proteins within the chloroplast. In theory it should be possible to deduce whether nuclear- or plastid-coded proteins are essential for chloroplast division and for cpDNA replication. In practice the results of such experiments are very difficult to interpret.

Leonard & Rose (1979) could detect no inhibition by CAP of plastid division or cpDNA synthesis over a period of 24 hours, using discs which had been cut from spinach leaves one day previously. However, neither could they detect any significant decrease in incorporation of ^{14}C-leucine into proteins of the chloroplasts, and chlorophyll synthesis was only partially inhibited. It is therefore possible that the failure to inhibit chloroplast division or cpDNA replication was caused by inadequate levels of CAP within the chloroplasts. When the experiment was repeated with 5-day-old, greening discs, CAP totally inhibited chlorophyll synthesis; with 6-day-old discs, chloroplast division was not inhibited by CAP. Since comparisons must be made between measurements on different ages of disc, these results cannot be regarded as conclusive; but they do suggest that continuous synthesis of proteins within the chloroplasts is not necessary for chloroplast division or cpDNA synthesis. Similar results have been reported for a moss, *Polytrichum* (Paolillo & Kass, 1977). Leonard & Rose (1979) also looked at the effects of CAP over a 48 hour period, and found that chloroplast division was inhibited, but cpDNA synthesis continued. This provides yet another example of the independence of chloroplast division and cpDNA replication. Aoshima *et al*. (1983) showed that CAP specifically inhibited the increase in level of 'chloroplastic' DNA polymerase in Chlorella. It would be particularly interesting to look for a similar effect in higher plants.

Complementary experiments to those described above were performed using cycloheximide. This was found to inhibit chloroplast division, cpDNA synthesis, cell growth, and radioactive labelling of both cytoplasmic and chloroplast protein fractions (Leonard & Rose, 1979). It has been suggested that the levels of the small subunit of ribulose bisphosphate carboxylase-oxygenase, synthesized in the cytoplasm may regulate the synthesis of the large subunit within the chloroplast (Barraclough & Ellis, 1979; Dean & Leech, 1982). If this interaction between cytoplasmic proteins and the synthesis of proteins in chloroplasts is widespread, it might be anticipated that inhibitors of cytoplasmic protein synthesis would also, indirectly, inhibit synthesis within the chloroplast. It is therefore not safe to conclude that processes inhibited by CHI require cytoplasmically synthesized proteins. Paolillo & Kass (1977) found effects in the moss *Polytrichum* similar to those reported for spinach discs. CHI specifically inhibited the production of an α-type, nuclear DNA polymerase in *Chlorella* (Aoshima *et al.*, 1983).

Given the difficulty of interpreting experiments involving inhibitors of protein synthesis, one can only surmise that a constant supply of cytoplasmically synthesized proteins might be needed for sustained division of chloroplasts and for cpDNA replication, and that proteins made in the chloroplast might not be involved in these processes, or perhaps remain active for a considerable period. If the results obtained with *Chlorella*, which showed inhibition of increases in 'chloroplastic' DNA polymerase levels by CAP are applicable to higher plants, one must assume that the γ-type DNA polymerase is relatively stable, and existing molecules are able to catalyse cpDNA replication in the presence of CAP. This is a question which clearly needs investigating.

Inhibitors of DNA synthesis

Experiments involving inhibitors of DNA synthesis have been performed to see if chloroplast division can continue in the absence of cpDNA replication. Again, the results of such experiments tend to be indicative rather than conclusive. Rose & Possingham (1976) were still able to detect chloroplast division in spinach discs and seedlings, after subjecting them to γ-irradiation at levels which were expected to eliminate all cpDNA replication. When Boasson & Laetsch (1969) treated tobacco discs with fluorodeoxyuridine (FUdR) to inhibit DNA replication, they found that kinetin-induced chloroplast division was inhibited, but light-induced division was not. Unfortunately, although it was shown that cpDNA synthesis occurs during recovery from FUdR treatment, no measurements were made of cpDNA synthesis during the treatment. Boasson & Laetsch suggested that, since chloroplasts contain many copies of their genome, several rounds of division should be possible in the absence of DNA synthesis. This does, of course, beg the question of why chloroplasts contain so many genome copies in the first place, if they can readily be diluted out by chloroplast division. By contrast, Rose *et al.* (1975) were able to inhibit plastid division in spinach discs by keeping them in darkness, but cpDNA synthesis continued. It therefore appears that chloroplasts can divide in the absence of DNA replication, and *vice versa*. This impression is supported by observations of the development of intact wheat and spinach leaves, as discussed earlier, in which cpDNA replication initially outstrips plastid division, causing an increase in the number of genome copies per plastid. Subsequently, plastid division is faster than DNA replication, so the cpDNA per plastid falls (Boffey & Leech, 1982; Ellis *et al*., 1983; Scott & Possingham, 1983).

THE MECHANISMS OF CHLOROPLAST DNA REPLICATION

In view of the apparent absence of coordination between chloroplast division and cpDNA synthesis, it is probably sensible to study these processes separately at present. The morphology of chloroplast division has been investigated in great detail, and has provided us with excellent evidence for the binary fission of both proplastids and more mature chloroplasts (Leech, 1976; Possingham, 1980; Leech, Thomson & Platt-Aloia, 1981). However, until a high level of synchrony of chloroplast division can be obtained in higher plants, it seems unlikely that the process will be amenable to biochemical investigation, especially since it probably depends on complex interactions between cytoplasm and organelle.

In 1980, Possingham stated 'There is almost no information available about the way in which the DNA of plastids is either replicated or segregated'. Four years later, the position has not changed dramatically, but there are indications that rapid progress may be possible in the near future. Our information is derived from two types of study: those which describe the behaviour of chloroplasts in intact tissue; and those which examine the characteristics of cell-free systems. In the former category are experiments in which intact leaves or leaf discs are incubated with ^{3}H-thymidine, and the sites and levels of incorporation of radioactivity are examined by autoradiography of thin sections of tissue, or by assays of radioactivity in the DNA of whole tissue and isolated chloroplasts. Such studies have indicated that cpDNA synthesis is carried out mainly by young chloroplasts and proplastids of wheat leaves, and it continues after nDNA replication has stopped (Boffey *et al*., 1979). It has also been shown that all the chloroplasts of expanding spinach leaf cells become labelled, rather than a sub-population of specialised replicating plastids (Rose *et al*., 1975). Owing to the effects of internal pools of non-radioactive thymidine, and the possibility of different rates of thymidine uptake by different cells and organelles, such labelling experiments are not well suited to precise measurements of rates of DNA replication. However, the results obtained with wheat have been shown to be valid by subsequent assays of DNA levels in different regions of the leaf (Boffey & Leech, 1982; Ellis *et al*., 1983).

Segregation of cpDNA

From microscopic examinations of chloroplast ultrastructure, it has become evident that cpDNA exists as clusters, or 'nucleoids', each containing several copies of the chloroplast genome. These nucleoids can be seen as fibrillar regions by electron microscopy, or as blue spots when isolated chloroplasts are viewed by ultra-violet fluorescence microscopy in the presence of the DNA-specific stain DAPI (4,6-diamidino-2-phenylindole). Such studies suggest that nucleoids are usually associated with plastid membranes, of either the thylakoid or inner envelope, and this attachment may be important in ensuring segregation of cpDNA to daughter plastids during division (Kowallik & Herrmann, 1974; Possingham & Rose, 1976; Rose, 1979; Sellden & Leech, 1981).

Replication origins

Electron microscopy has been of use in providing information about the mechanism of cpDNA replication. D-loops have been observed in the cpDNA of pea and maize, apparently initiated about 7kb apart from each other (Tewari, Kolodner & Dobkin, 1976). These loops increase in size as DNA replication proceeds, extending towards, and eventually through, each other to give a Cairns-type of replicative intermediate (Fig. 1). Rolling circle intermediates have also been seen, and it is thought that these result from the initiation of a second round of replication immediately after the first. Cairns-type replicative intermediates have also been seen in Euglena (Ravel-Chapuis, Heizman & Nigon, 1982), and in this case the replication origin has been mapped near the 5' end of the supplementary 16S rRNA gene. This mapping was not precise enough to rule out the possibility that there may be two replication origins near each other, as in pea and maize.

A different approach to the identification of replication origins is that described by Stinchcomb et al. (1980). The plasmid YIp5 contains a prokaryotic origin of replication, and the yeast gene ura3. It can therefore be cloned in bacterial cells, but will not transform yeast ura3 deletion mutants, since it does not contain a eukaryotic replication origin, and will therefore not be replicated within the yeast cells. If a fragment of DNA containing a eukaryotic origin of replication is inserted into YIp5, the plasmid will be replicated within yeast cells, and can be detected by its ability to transform ura3 deletion mutants. Such inserted sequences are known as 'autonomously replicating sequences', or ars fragments. Using this system, Zakian (1981) isolated an ars fragment, 2.2 kb in length, from the mitochondrial DNA of *Xenopus*, demonstrating that it is not only nuclear DNA which contains eukaryotic replication origins. A particularly exciting result using YIp5 has been obtained by Uchimiya *et al.* (1983), who isolated an *ars* fragment 1.7kb long from tobacco chloroplast DNA, produced by restriction with EcoRI. This fragment apparently maps in the smaller, single-copy region of the cpDNA, between the inverted repeats. Since the rate of cpDNA replication is possibly determined by the frequency of its initiation, the isolation of cpDNA replication origins is likely to be of enormous value in identifying factors which interact with the origin, and which might therefore be involved in the regulation of cpDNA replication.

DNA polymerases

Little is known about the enzymes involved in cpDNA replication. Sala *et al.* (1980, 1981) isolated a γ-type DNA polymerase from the chloro-

Fig. 1. The D-loop mechanism for replication of chloroplast DNA. Note that synthesis of the two daughter strands is asynchronous. 1 = origin for synthesis of the first daughter strand; 2 = origin for synthesis of second daughter strand.

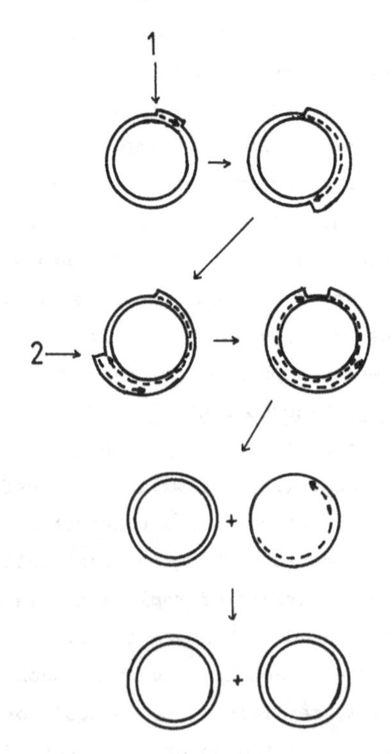

plasts of rice and spinach. It was shown to have a molecular weight of about 105,000, and to have a requirement for Mn^{2+} and high ionic strength, with a pH optimum of 8-9. *In vitro* it was inhibited by N-ethylmaleimide, but was completely resistant to aphidicolin, which was a strong inhibitor of the (nuclear) α-polymerase (see Dunham & Bryant, this volume). Autoradiography showed that, *in vivo*, aphidicolin inhibited the synthesis of nDNA but not that of organelle DNA, suggesting that the γ-type polymerase does actually play a part in the synthesis of cpDNA. Aoshima *et al*. (1983) measured DNA polymerase activities at various stages of the cell cycle of synchronously dividing *Chlorella*, and detected two different activities, one of which increased during nDNA replication, the other rising when cpDNA replication began. In this case the 'chloroplast' polymerase did not seem to have a requirement for Mn^{2+}.

Cell-free systems

The problems of interpreting the results of experiments performed on intact tissue have already been discussed, and it seems clear that progress in understanding the biochemistry of cpDNA replication will depend to a great extent on the use of cell-free systems.

Tewari *et al*. (1976) were able to solubilize the DNA synthesizing activities of pea chloroplasts, and they showed that this soluble preparation would catalyse semi-conservative replication of DNA, using cpDNA as a template. Chromatography in a DEAE-cellulose column resulted in the separation of three DNA synthesizing fractions, which differed in their specificities for single- or double-stranded DNA as template.

Bohnert, Schmitt & Herrmann (1974) obtained light-driven incorporation of ^{3}H-thymidine into the DNA of chloroplasts isolated from spinach leaves. No Mg^{2+} was needed for this incorporation, which occurred only with intact chloroplasts. Incorporation of label into DNA could also be achieved by broken chloroplasts, but only in the presence of deoxyribonucleoside triphosphates, using ^{3}H-thymidine triphosphate as the source of label; this incorporation was not light-driven. It was not shown whether the incorporation of label into DNA was the result of replication or repair of the DNA. The levels of incorporation were relatively low in this system, which may explain why it does not appear to have been used for further investigations of cpDNA replication.

Recently, Mills & Baumgartner (1983) reported the light-driven incorporation of ^{3}H-thymidine into the DNA of chloroplasts isolated from pea leaves. The highest rates of synthesis were obtained using chloroplasts isolated from young leaf tissue, giving about 500,000 cpm/mg chlorophyll over 45 min, which is an enormous improvement on the levels of incorporation reported by Bohnert *et al.* (1974). Mills & Baumgartner attribute their success to the use of young tissue, in which the chloroplasts are naturally synthesizing DNA at a high rate. It is not known if this incorporation is the result of repair or replication of DNA. By isolating chloroplasts from wheat under conditions known to produce functionally intact organelles (Leegood & Walker, 1979), we have been able to obtain light-driven incorporation of ^{3}H-thymidine into cpDNA, at levels of about 100,000 cpm/mg chlorophyll, over a period of 30 min (M. Jones & S.A. Boffey, unpublished data). Like Mills & Baumgartner, we find this incorporation is at its highest in chloroplasts isolated from young tissue. Although it has not been shown directly in either the pea or wheat systems that it is chloroplast DNA which is being labelled, light-stimulation of the process can be taken as strong evidence that it is taking place within the chloroplast (Ellis & Hartley, 1982), and the need for intact chloroplasts supports this conclusion.

It is obviously important to provide conclusive evidence for the identity of the labelled DNA, and to show that it is a consequence of replication, rather than repair, before the isolated chloroplast systems can be used for further studies. However, it seems likely that they will open up the possibility of detailed biochemical studies of the replication of cpDNA in higher plants, with the prospect of identifying those factors which regulate replication in the intact plant.

REFERENCES

Aoshima, J., Nishimura, T. & Iwamura, T. (1983). DNA polymerases of *Chlorella*. I. Chloroplastic and nuclear DNA polymerases in synchronized algal cells. *Cell Structure and Function*, 7, 327-40.

Barraclough, R. & Ellis, R.J. (1979). The biosynthesis of ribulose bisphosphate carboxylase. Uncoupling of the synthesis of the large and small subunits in isolated soy bean leaf cells. *European Journal of Biochemistry*, 94, 165-77.

Bennett, J. & Radcliffe, C. (1975). Plastid DNA replication and plastid division in the garden pea. *FEBS Letters*, 56, 222-5.

Boasson, R. & Laetsch, W.M. (1969). Chloroplast replication and growth in tobacco. *Science*, 166, 749-51.

Boffey, S.A., Ellis, J.R., Sellden, G. & Leech, R.M. (1979). Chloroplast division and DNA synthesis in light-grown wheat leaves. *Plant Physiology*, 64, 502-5.

Boffey, S.A. & Leech, R.M. (1982). Chloroplast DNA levels and the control of chloroplast division in light-grown wheat leaves. *Plant Physiology*, 69, 1387-91.

Bohnert, H.J., Schmitt, J.M. & Herrmann, R.G. (1974). Structural and functional aspects of the plastome. III. DNA- and RNA-synthesis in isolated chloroplasts. Portugaliae Acta Biologica, 14, 71-90.

Cattolico, R.A. (1978). Variation in plastid number. Effect on chloroplast and nuclear deoxyribonucleic acid complements in the unicellular alga Olisthodiscus luteus. Plant Physiology, 62, 558-62.

Dean, C. & Leech, R.M. (1982). The co-ordinated synthesis of the sub-units of ribulose bisphosphate carboxylase in a wheat line with alien cytoplasm. FEBS Letters, 144, 154-6.

Ellis, J.R., Jellings, A.J. & Leech, R.M. (1983). Nuclear DNA content and the control of chloroplast replication in wheat leaves. Planta, 157, 376-80.

Ellis, R.J. & Hartley, M.R. (1982). Preparation of higher plant chloroplasts active in protein and RNA synthesis. In Methods in chloroplast molecular biology, ed. M. Edelman, R.B. Hallick & N.-H. Chua. pp. 169-88. Amsterdam: Elsevier.

Kowallik, K.V. & Herrmann, R.G. (1974). Structural and functional aspects of the plastome. II. DNA regions during plastid development. Portugaliae Acta Biologica, 14, 111-26.

Lamppa, G.K. & Bendich, A.J. (1979). Changes in chloroplast DNA levels during development of pea (Pisum sativum). Plant Physiology, 64, 126-30.

Leech, R.M. (1976). The replication of plastids in higher plants. In Cell Division in Higher Plants, ed. M.M.Yeoman. pp. 135-59. London and New York: Academic Press.

Leech, R.M., Thomson, W.W. & Platt-Aloia, K.A. (1981). Observations on the mechanism of chloroplast division in higher plants. New Phytologist, 87, 1-9.

Leegood, R.C. & Walker, D.A. (1979). Isolation of protoplasts and chloroplasts from flag leaves of Triticum aestivum L. Plant Physiology 63, 1212-4.

Leonard, J.M. & Rose, R.J. (1979). Sensitivity of the chloroplast division cycle to chloramphenicol and cycloheximide in cultured spinach leaves. Plant Science Letters, 14, 159-67.

Mills, W.R. & Baumgartner, B.J. (1983). Light-driven DNA biosynthesis in isolated pea chloroplasts. FEBS Letters, 163, 124-7.

Paolillo, D.J. & Kass, L.B. (1977). The relationship between cell size and chloroplast number in the spores of a moss. Polytrichum. Journal of Experimental Botany, 28, 457-67.

Possingham, J.V. (1973a). Effect of light quality on chloroplast replication in spinach. Journal of Experimental Botany, 24, 1247-60.

Possingham, J.V. (1973b). Chloroplast growth and division during the greening of spinach leaf discs. Nature New Biology, 245, 93-4.

Possingham, J.V. (1980). Plastid replication and development in the life cycle cycle of higher plants. Annual Review of Plant Physiology, 31, 113-29.

Possingham, J.V. & Rose, R.J. (1976). Studies of the growth and replication of spinach chloroplasts and of the location and segregation of their DNA. In Genetics and Biogenesis of Chloroplasts and Mitochondria, ed. T. Bucher, W. Neupert, W. Sebald & S. Werner. pp. 387-90. Amsterdam: Elsevier.

Possingham, J.V. & Smith, J.W. (1972). Factors affecting chloroplast replication in spinach. Journal of Experimental Botany, 23, 1050-9.

Ravel-Chapuis, P., Heizmann, P. & Nigon, V. (1982). Electron microscopic localization of the replication orogin of Euglena gracilis chloroplast DNA. Nature, 300, 78-81.

Rose, R.J. (1979). The association of chloroplast DNA with photosynthetic membrane vesicles from spinach chloroplasts. Journal of Cell Science, 36, 169-83.

Rose, R.J., Cran, D.G. & Possingham, J.V. (1975). Changes in DNA synthesis during cell growth and chloroplast replication in greening spinach leaf disks. Journal of Cell Science, 12, 27-41.

Rose, R. & Possingham, J. (1976). Chloroplast growth and replication in germinating spinach cotyledons following massive γ-irradiation of the seed. Plant Physiology, 57, 41-6.

Sala, F., Parisi, B., Burroni, D., Amileni, A.R., Pedrali-Boy, G. & Spadari, S. (1980). Specific and reversible inhibition by aphidicolin of the α-like DNA polymerase of plant cells. FEBS Letters, 117, 93-8.

Sala, F., Galli, M.G., Levi, M., Burroni, D., Parisi, B., Pedrali-Noy, G. & Spadari, S. (1981). Functional roles of the plant α-like and γ-like DNA polymerases. FEBS Letters, 124, 112-8.

Scott, N.S. & Possingham, J.V. (1983). Changes in chloroplast DNA levels during growth of spinach leaves. Journal of Experimental Botany, 34, 1756-67.

Sellden, G. & Leech, R.M. (1981). Localization of DNA in mature and young wheat chloroplasts using the fluorescent probe 4'-6-diamidino-2-phenylindole. Plant Physiology, 68, 631-4.

Stinchcomb, D.T., Thomas, M., Kelly, J., Selker, E. & Davis, R.W. (1980). Eukaryotic DNA segments capable of autonomous replication in yeast. Proceedings of the National Academy of Science of the U.S.A., 77, 4559-63.

Tewari, K.K., Kolodner, R.D. & Dobkin, W. (1976). Replication of circular chloroplast DNA. In Genetics and Biogenesis of Chloroplasts and Mitochondria. ed. T. Bucher, W. Neupert, W. Sebald & S. Werner. pp. 379-86. Amsterdam: Elsevier.

Tymms, M.J., Scott, N.S. & Possingham, J.V. (1983). DNA content of Beta vulgaris chloroplasts during leaf cell expansion. Plant Physiology, 71, 785-8.

Uchimiya, H., Ohtani, T., Ohgawara, T., Harada, H., Sugita, M. & Sugiura, M. (1983). Molecular cloning of tobacco chromosomal and chloroplast DNA segments capable of replication in yeast. Molecular and General Genetics, 192, 1-4.

Whiteway, M.S. & Lee, R.W. (1977). Chloroplast DNA content increases with nuclear ploidy in Chlamydomonas. Molecular and General Genetics 157, 11-15.

Zakian, V.A. (1981). Origin of replication from Xenopus laevis mitochondrial DNA promotes high-frequency transformation of yeast. Proceedings of the National Academy of Science of the U.S.A., 78, 3128-32.

INDEX

SOCIETY FOR EXPERIMENTAL BIOLOGY SEMINAR SERIES

A series of multi-author volumes developed from seminars held by the Society for Experimental Biology.
Each volume serves not only as an introductory review on a specific topic, but also introduces the reader to experimental evidence to support the theories and principles discussed, and points the way to new research.

1. Effects of air pollution on plants
Edited by T.A. Mansfield

2. Effects of pollutants on aquatic organisms
Edited by A.P.M. Lockwood

3. Analytical and quantitative methods in microscopy
Edited by G.A. Meek and H.Y. Elder

4. Isolation of plant growth substances
Edited by J.R. Hillman

5. Aspects of animal movement
Edited by H.Y. Elder and E.R. Trueman

6. Neurones without impulses: their significance for vertebrate and invertebrate systems
Edited by A. Roberts and B.M.H. Bush

7. Development and specialisation of skeletal muscle
Edited by D.F. Goldspink

8. Stomatal physiology
Edited by P.G.Jarvis and T.A.Mansfield

9. Brain mechanisms of behaviour in lower vertebrates
Edited by P.R. Laming

10. The cell cycle
Edited by P.C.L. John

11. Effects of disease on the physiology of the growing plant
Edited by P.G. Ayres

12. Biology of the chemotactic response
Edited by J.M.Lackie and P.C.Williamson

13. Animal migration
Edited by D.J. Aidley

14. Biological timekeeping
Edited by J. Brady

15. The nucleolus
Edited by E.G. Jordon and C.A. Cullis

16. Gills
Edited by D.F. Houlihan, J.C. Rankin and T.J. Shuttleworth

17. Cellular acclimatisation to environmental change
Edited by A.R. Cossins and P.Sheterline

18. Plant biotechnology
Edited by S.H. Mantell and H. Smith

19. Storage carbohydrates in vascular plants
Edited by D.H. Lewis

20. The physiology and biochemistry of plant respiration
Edited by J.M. Palmer

21. Chloroplast biogenesis
Edited by R.J. Ellis

22. Instrumentation for environmental physiology
Edited by B. Marshall and F.I. Woodward

23. The biosynthesis and metabolism of plant hormones
Edited by A. Crozier and J.R. Hillman

24. Coordination of motor behaviour
Edited by B.M.H. Bush and F. Clarac

25. Cell ageing and cell death
Edited by I. Davies and D.C. Sigee

26. The cell division cycle in plants
Edited by J.A. Bryant and D. Francis